NO SHORTCUTS

NO SHORTCUTS

One Woman's Journey from Startup to Successful
Sale of a Multimillion-Dollar Tech Company

GAIL PEACE

Gp

Motivating
Entreprenuers

Gail's books may be purchased for educational, business, or sales promotional use. For information please contact: info@thatgailpeace.com

Production and creative provided by Epiphany Creative Services
Cover design and art by Mike Quinones G
Printed in the United States of America
Library of Congress Cataloging-in-Publication Data
Library of Congress Control Number: 2025903959

FIRST EDITION

Gail Peace – 1st ed.
TITLE: No Shortcuts: One Woman's Journey from Startup to Successful Sale of a Multimillion-Dollar Tech Company
p. cm.

ISBN: 979-8-9925436-0-5 (Paperback)
ISBN: 979-8-9925436-1-2 (eBook)

1. BUS025000 BUSINESS & ECONOMICS / Entrepreneurship
2. BUS109000 BUSINESS & ECONOMICS / Women in Business

Distributed by That GP Press

14 10 9 8 7 6 5 4 3 2

To my family—my mother and father,
who gave me the gift of dreams and the ability
to realize them, and my brother,
who has always stood by me with unwavering belief.

Contents

Chapter 1

What Dumb-A*# MBA Thought of This?

As I sit looking out the window of my dream house, admiring the lake view, the trees, and the golf course, I'm overcome with an immense feeling of gratitude and—dare I say it—pride. Located right outside Nashville, the new "it" city, my upscale, private community is a very coveted neighborhood. I'm in a good place, and I'm fortunate to have just sold the very company I hastily sketched out at a red picnic table and built from the ground up. I had a dream to build a Software as a Service, or SaaS, platform and ultimately sell it. Life is good. Very good.

The road to where I am now has not been one of ease or speed. It has had many twists and turns. Some funny, some sad, some exasperating. But all necessary to get me to this place. As I sat down to write this book, the most poignant memories immediately came to mind. I asked myself over and over where I should start, and then it hit me. What better place to begin than with me sitting in an exam room surrounded by physicians who were surprised to discover—they were my clients!

Medical Moment

I had just turned fifty. Hitting the half-century mark, as a rite of passage, brings along with it certain milestones and protocols. One of those is the obligatory colonoscopy. Lucky me.

As fate would have it, my primary care physician recommended a doctor in a hospital system that happened to be one of my favorites in the Nashville area. It also happened to be one that utilized the medical app and software I had created to help medical facilities and doctors track their time in order to get paid correctly by the hospital.

On appointment day, I was sitting in the waiting room filling out the stack of new patient paperwork when it dawned on me that this physician may in fact be using the very software I had created. I couldn't wait to find out if he was a user of our software.

When he walked into the exam room, he paused for a moment and said, "Gosh, you look familiar."

"Maybe it's because I am famous," I said.

From what I have been told over the years, I have one of those faces. You know, the kind where everybody feels they know you but just can't place how or where. I couldn't wait to reveal my big secret. I seized the moment and succinctly explained that I was the founder of Ludi and had created the software DocTime®. I asked him if he was familiar with it.

"I am! How cool is that? I used to have to turn in these paper time logs, and I would always lose them through the month and have to start over. I wasted so much of my time. Now I use this little app throughout the month, and it is super easy," he said.

We began to chat a bit, and he mentioned that ironically, he'd recently shown a colleague how to use the app. It was another physician who had received a new contract with the hospital.

"Wait a minute," he said. He jumped up, headed out the door, and disappeared into the hallway.

When he came back in, he had another physician with him. It was official. My physician visit had turned into a DocTime client meeting. It doesn't get much better than that. This was a full-circle moment for me. You just can't write this stuff. Or can you?

I had started Ludi, my software company, six and a half years prior to that moment on the exam table. I had worked so hard and spent so much time growing Ludi into a full-fledged business with a team of people who could work with clients in my place that it had been years since I had been out in the field directly with physicians. Now, as I sat at this appointment, I was hit with the realization that all those long hours, all that stress, everything it had cost me physically and emotionally—it had been worth it.

Tell Me Where It Hurts?

I took a brief second to appreciate the scene and savor the moment. But not for long. I instinctively jumped into client service mode. I asked them what they liked about the app and what they didn't like. My doctor had to switch between agreements. He had been using the software for years and said he could be more efficient if he didn't have to log out and log back in to move to another contract.

"Oh, you're right and you don't have to. You just hit this little button here!"

"Aah," he exclaimed, his tone a bit embarrassed.

My doctor listed a few other concerns my team had heard before. One issue was easily solved with a couple of quick user interface changes in the settings that were already available. Happily for me, his other feedback was more of a function of the regulatory world.

Standing across from me, the physician with the new teaching agreement continued to grow excited that he was now getting paid by the hospital for this function. He had been interested in teaching for a long time and was now part of the faculty. He was very proud and did not seem at all bothered the hospital asked him to keep track of his hours in order to receive this pay.

I told them both that we were in the process of adding biometric login so they would soon be able to open the app with a fingerprint or their face, which they were both clearly excited about.

They left the room, but my mind was still in a blur from what had just transpired. My routine medical visit had turned into an impromptu feedback session with my direct end users, and the data that was coming in was good. Very good.

Diagnosis, Please?

As the creator of a company and designer of a technical app, I wanted to know: Did it work? Did it solve the intended problem? Did the work flow work for the users? Were my efforts in vain, or is this relevant and needed?

It was a relief to hear they found the software to be easy to use and saved them time. They made it very clear they didn't like keeping track of this information on paper as they had to do in the past. My doctor said it saved him time on the days he worked on his medical directorship. He recorded the time through the month as he completed it. In the past, he had to sit down at the end of the month with his calendar and try to remember what he had done in order to fill out the paper time log. Not anymore. Ludi's app was the answer. Now, with the app on his phone, he could handle all of that at the end of a meeting, when leaving the classroom, or at the end of a week with his students.

On the day of my actual colonoscopy, my doctor poked his head into my prep room to give me the rundown and ease my anxieties about the upcoming anesthesia. After politely answering my questions, he shifted and said, "Hey, I like the new app, and I even got the biometric login going. I also like how I can see the detail behind each duty by clicking on this detail button."

I smiled to myself with the knowledge that he could have done that all along! He just didn't know. It was a thrill to have had that opportunity to see firsthand how a product I'd created, in my own little brain, had made a physician's life easier—in real time. What a rush!

Beyond the Horizon

Shortly thereafter, I decided I needed a vacation. I was visiting friends in Chicago and learned of a pending trip they were taking to Iceland. The couple was taking her sister along, so there were three of them embarking on a trip of a lifetime. I hastily suggested I be her roommate on the trip. Then I thought, *Gosh, that was rude*, so I didn't say anything else. The following week, my friend called and said they would be thrilled if I joined them. Everything had been planned. All I had to do was RSVP and write a check. It was an amazing offer, and I jumped at the chance to go. Iceland is an amazing country. If you haven't been, I highly recommend you go if you are ever afforded the opportunity.

Once we were all settled in and the itinerary began to unfold, I found myself joining in and tagging along on all the planned activities. My favorite was the Blue Lagoon. The natural hot springs that exist in Iceland are amazing. Water is heated by volcanic activity below two tectonic plates meeting deep below the earth's surface and bubbles up to the surface. The Blue Lagoon is an enormous hot spring where you can get a reservation, don your swimming suit, and relax for hours moving around the spring, where drinks, facial masks, and other treats are offered.

The Blue Lagoon was created in 1976 from a pool that formed next to a geothermal power plant. The power plant drills for boiling water, which is taken from the ground and used to heat fresh water, then pumped through radiators to heat Iceland's houses, businesses, even streets. The water pulled from deep inside the earth is 200 degrees Celsius (392 degrees Fahrenheit) and is milky from dissolved minerals mixed in with sea water. This water was released into a nearby lava field that created a silica layer and then pooled with the rich mineral water. The spa was created taking advantage of this rich mineral geothermal sea water that continues to be released from the power plant today.

One of the more active trips of our adventure was climbing into an ice cave in a glacier. Glaciers have caves that are created each summer as the glacier melts. Now, let me say first, I tend to suffer a bit from claustrophobia. Probably not a good match for this particular jaunt.

As we entered the cave, I could sense a bit of panic bubbling deep down inside. The small spaces seemed to be closing in, and I wasn't able to calm my breathing. I had to be escorted out of the cave by the guide, leaving the others momentarily as he got me safely out. The guide left me outside this particular cave, on the edge of the glacier. Looking out over the vast, incredible terrain, I realized I had never experienced anything quite like this phenomenon before. As far as I could see in any direction, there were no people and no indications of civilization; it was 100 percent nature. Fans of *Game of Thrones* well know that the show was filmed here. It's simply stunning and a very unique vista.

Quiet. Calming. No background noise. No one texting or emailing or calling. I couldn't even see the van we had arrived in, so I knew I was totally isolated in nature. It was breathtaking. It really made me feel like, *Hey, you really are not that big a deal; look at all this nature around you.*

I was enjoying the time alone with my thoughts when it happened—my cell phone rang. I was stunned. I thought maybe it was God calling me directly.

"Hello?" I said.

"Gail? This is Dr. Finn. I have a question."

Dr. Henry Finn was one of the very first physician users of my DocTime app. The call was coming in not on cellular service but via the internet. I was completely taken aback that I was connected to a Wi-Fi network. *The irony! I'm in the middle of nowhere, and a physician using the software I created is calling me!*

He proceeded to explain an issue he was having with a particular entry he'd made. As my connection was precarious, I texted a colleague asking her to reach out to him and assist with his issue. We had been in business for five years, and I loved the fact that physicians still called me directly, which is why I picked up the phone three thousand miles and several oceans away.

Blow Hard from the Windy City

There was one physician who was also part of our early story. After the software was live for one full month at his hospital, he was the only physician who had not yet logged in to the software. This particular client was four miles from my home, so it was easy to pop over and politely stalk him. I was waiting for him outside of his weekly surgical meeting. Dr. Jones was very frustrated having to keep track of his time.

"Hey, Dr. Jones," I said in as friendly a tone as I could muster. I had worked at this hospital and knew all the physicians well. "Do you have five minutes so I can show you how you can now log your time for your teaching agreement and medical directorship contracts electronically?"

He reluctantly agreed, so I set up the app on his phone; I had already reset his password. It literally took three minutes. While doing this, I explained that his hospital was asking him to track his time electronically for his teaching and co-management agreements. He looked me in the eye and said, "What dumb-ass MBA student invented this?"

I looked right back at him in the eye and said, "Why, as a matter of fact, it was me!"

I told him I had left the hospital five months prior when my job was eliminated and had built this software and app. Without missing a beat, he said, "Why, that seems like a good idea, and it will be fine to use."

I found myself slipping into founder mode. I provided a thirty-second walk-through of the Stark Law and the regulatory environment. The answer boiled down to a simple truth: the government is the largest consumer of health care through the Medicare and Medicaid system, so physician compensation must be set at fair market value and carefully monitored. Otherwise, these costs can unnecessarily drive up the cost of care to the largest consumer—the government. There are rules that must be followed: the duties have to be specified, the pay has to be defined in advance and be at fair market value, time logs must be obtained as proof of service, etc. This scenario can create a liability for the hospital as well as the physician.

I realized it was turning from conversation to pretty complicated discussion rather fast. So, to bring it on home, I mentioned that the software solution I designed protects the hospital and the physician. Once I explained that small but important point, he understood better why they had to record their time. Issue one was solved.

Dr. Finn, on the other hand, who rang me in Iceland, was a critical contributor to the early days of my company. I had known and admired him for years, a very savvy businessman in his own right. He called me one evening after we had gone live with the software at his hospital. He was so complimentary of me having had the guts to start the company, and he thought the idea was really solid.

He even offered me investment money, which was the biggest compliment ever. I mean, Dr. Henry Finn is one of the top physicians in orthopedics in not only Chicago but the country! He already had multiple inventions at that time under his hat and liked what he saw in mine. While I didn't accept his offer, he helped me have the confidence in those early days that, yes, this product was needed, physicians would use it, and it met a need!

So You Have an Idea, Now What?

As I think back to my early path to starting Ludi and what the journey entailed to figure out how to start a company, build a software product, finance it, sell it, deliver it, hire people, it was a smorgasbord of all new things to figure out. Many people have ideas for products every day, so what do you do when you have one? Sure, there are the mechanics of building a business, for example, building a business plan, testing your idea, confirming there is a market, etc., but under all that, what makes people ultimately take the plunge?

Why do some people follow through with chasing their dream to build something on their own while others don't? Change is always hard, and often it is done only when the pain of making a change is less than the pain of staying with what one is currently doing. If people find themselves in a job they don't like, most would just get a new job, so that doesn't seem to be the answer to describe what motivates people to chase a dream. Why do some people break out on their own, taking huge personal and financial risk? This led me to another question: are entrepreneurs born or created? Nature versus nurture, the age-old question.

Is there a certain quality people are born with that makes it easier for them to make the decision to start a business? Or is it specialized training that helps people define a path forward? Maybe it is somehow a combination.

Trail Markers:

- *Your product has arrived when you see clients using it in their day-to-day job, and it is the greatest source of pride in a job well done.*
- *The way clients use your product may be in fact a bit different than envisioned; keep iterating with feedback from users.*
- *Sometimes the toughest to win over in adopting technology will become your best advocates and references for future clients.*

Chapter 2

It's All in How You're Wired

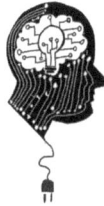

I have come to believe the catalyst, at the root, for taking the risk to start your own company is somewhat how you are born, the drive inside you. When I was a young girl, my mom took me to a cake decorating class at the local grocery store. I remember enjoying mixing the icing to the exact correct consistency to flow nicely from the bag. I loved it.

Soon afterward, I was attending a birthday party and saw a Barbie doll cake. Everyone those days had parties with themed cakes. I looked at it and thought, I *could make that!* An idea emerged that I could make birthday cakes for the neighborhood kids. I decided that I would need to get a Tony the Tiger mold (for boys) and a Barbie doll mold (for girls). I would need some specific tips for the icing bags, all told, probably a ten-dollar investment—remember, it was 1977! I prepared for my first "angel investor" pitch, to my parents.

"Okay, so the Barbie doll cake is eighteen dollars at the store," I began. "The cake molds are seven dollars, the tips are two dollars, and I figure I will need one dollar for the ingredients per cake." I have no idea what my parents were thinking, but I chattered on, my energy commanding the moment. "If you'll buy the supplies and the ingredients, then I'll pay you one dollar per cake for each one I sell along with the cost of the supplies until I pay you back!"

I had done the math, and I had a business plan. I figured I could sell the cakes for twelve to eighteen dollars, so it seemed like a great strategy to me. I'd be able to pay off the investment after selling ten cakes. I recall a surprised look on my parents' faces. I expect they were wondering how I had come up with this business plan.

My parents had encouraged us to save the money we earned from odd jobs. These lessons served my brother and me well. I am not sure many kids we knew back then had the opportunities we had to learn about money. People didn't tend to educate their children about finances in those days, and I'm not sure it's much better today.

I Was Schooled

We had moved to a new community with much more rigorous schools. My brother and I both had a bit of a time adjusting. The secondary school was made up of primarily children of Jewish descent. On the Jewish holidays, of the ninety or so kids in my grade, there would only be three of us present: Louis Papas, Kris Jones, and myself! I always begged my mom to let me stay home, but that was never approved; she didn't believe there would be only three of us. But as it turned out, those were some fun school days.

We had moved from the north side of the suburbs to the west side, and the schools were harder. There was a ton of pressure on us kids to excel in school. It was cool to be smart. I had always done well in school, so I fell into the studious category. But I am not going to lie, this school kicked my booty. And I was a bit behind from attending schools on the north side.

Junior high was where I truly learned how to study. I credit that school for my ability to do well all the way through college. They trained us that well. Honestly, doing well during that time was harder for me than all the following years thereafter. But it wasn't perfect by any means. There was a noticeable undercurrent of tension between the haves and the have-nots. The school was in an extremely affluent neighborhood.

While we lived in a middle-class neighborhood, we were surrounded by those on the higher end of prosperity in surrounding neighborhoods. We were the family not running the air conditioner when it was ninety-five degrees.

I remember vividly beginning seventh grade. Many of my classmates had fancy designer shoes and jeans. I did not. My mom sewed many of my clothes, and they were fine, but they didn't have the designer labels. This was where I learned about fancy purses, jeans, and above all, Nikes.

Nikes were the rage. I remember saying to my mom, "I need to wear Nikes to fit in." Her response was, "Those shoes are forty to fifty dollars, and we don't spend that on tennis shoes. We spend twenty." I was embarrassed that I didn't have Nikes and fancy jeans, so I wanted to earn money so I could fit in to a certain extent. In those days, some kids may have had an allowance, but we were not given money by our parents. They provided our clothing and met our needs, but if we wanted something special, e.g., expensive Nikes, we were expected to earn the money.

I think this is why my brother and I started working so early, certainly not in the salt mines but doing various jobs for spending money. In other words, we were not breaking any child labor laws, but we were entrepreneurial in our ability to earn. Luckily, my dance card was not filled in those junior high awkward years.

While I make friends early, I was not early in the dating scene. I had the time and I had the drive to earn money.

First Impressions

One of my favorite jobs was delivering pizzas by boat one summer. I loved it. Then there was the year I worked for a small private cruise line. (The kind you see on the reality TV show *Below Deck*. That show is more real than you can imagine!) I stretched that stint to fifteen months and then headed off to graduate school.

My family, especially my dad, valued education. He even strongly encouraged my mom to go back to school to get a college degree after my brother and I were born, with this reasoning: What will you do if something happens to me? You have to be able to take care of these kids.

I had a very deliberate, well thought out strategy for success in my classes. I stayed very quiet, studied hard, raised my hand, and was over the top when it came time to studying for the first test. I was working on my master's degree in health administration and also decided I wanted to balance it with a master's degree in business administration, with an emphasis in finance. I was pretty sure I wanted to be in health care but thought if I changed my mind in the future, having the straight business degree might be helpful. The two programs were in fact separate but overlapped in many of the core classes.

Each degree was sixty hours, so getting both totaled seventy-two hours with the classes that overlapped. If I lined up all the core and electives just right, the plan was to get all my credits done within two years. Seventy-two hours divided by four meant I needed to take eighteen hours each semester or have two semesters with eighteen and two with fifteen, and six hours in the summer.

In graduate school, twelve to fifteen hours is considered a full load, but I knew I could do it. For heaven's sake, I was no longer scrubbing the galley down with Mikro-Quat. Working for the cruise line was an incredible physical feat. We worked all the time. I still have scars on my hands from using this chemical, Micro-Quat, which was later deemed toxic by the World Health Organization. But it killed bacteria—and more, apparently! Anyway, school wasn't going to be harder than all the physical work I had been doing for the past fifteen months.

I turned in my class selections and learned that I had to have the dean approve my schedule because I wanted to take eighteen hours. I made an appointment with the dean of the medical school, as the MHA program was part of the medical school. (Talk about intimidating.)

When we met, he asked why I was taking so many hours. I explained that two years was most comfortable financially for me, and too, with what I had saved it meant that I wouldn't be more of an imposition on my parents. I told him I was used to physically working fifteen to sixteen hours a day. I could certainly do this. This was a different kind of work obviously, but it was still something I would approach with my solid work ethic. He then explained that if I didn't get at least a 3.0, he would not approve it again. Imagine his surprise when I arrived the next semester with a 4.0!

Our accounting class had the first test across both the MHA and the MBA programs. I studied hard, but I am fortunate that my brain is pretty numbers oriented, so it was the best first test for me to have.

On the day the test was returned, the professor wrote the scores on the board. She then turned to the class and announced, "It is a really good thing I grade on a curve, because if I didn't only one person in here would have passed."

I remember thinking, *Gosh, it was a hard test, but heavens, I don't think I did that bad.* I started to get a bit nervous.

She wrote a 92 on the board, then below it a 69, 67, 63, etc. Then she walked to hand out the tests and tossed mine onto my desk, 92! That's all it took. From that day forward I was branded as smart. Not only by my peers but also by the professors.

My strategy had worked. I worked hard, kept quiet the first month, and by doing so had set the stage for my success. But I also began to learn something about biases.

My professors began assuming I would do well, so it almost became a sort of self-fulfilling prophecy. For example, if they were grading an essay I had written, did my past scores influence the future score? Do professors have a bias as the students become established? Probably.

It is really hard to change biases, so yes, I believe their perceptions did influence my future scores. It seems to be human nature and a part of life to take experiences and apply them forward. I would soon come to learn that lesson in an even deeper way as I moved into my career. First impressions matter; if you appear to be the smartest in the room, others will see you that way.

Tip 1: First Impressions

My advice when starting at a new school or in a new job is to shut your mouth for the first two weeks and observe. Here is my formula:

1. Keep your mouth closed for two weeks, do not ask a lot of questions, and observe. Things will become clear as you learn the ropes.

2. Write down what you don't know, and research it on your own each evening on the googlies! If you are in school, study your behind off for your first tests.

3. Get to know your colleagues by listening to them; share as little as possible while you determine who you can trust. Those first impressions will carry you throughout your relationships.

Putting It into Practice

I was recruited to the Sachs Group when I was twenty-seven. This company was my ticket to Chicago! Not that Southeast Michigan wasn't for me, but well, it was a difficult place to live as a single woman who was not from there.

When I took the job, I looked at apartments, which proved to be tricky. I drove from Southeast Michigan and stayed at a hotel in Chicago to begin my search for an apartment on the north side. There was very little inventory, so not much to view. What I did find was old, grungy, and dirty. This was all pretty overwhelming for a girl who had grown up in the suburbs of St. Louis and then lived in the suburbs of Detroit, but I figured it out. I decided after one day, apartments were not going to work for me.

I called my mom, who was at that time a real estate agent in St. Louis. I asked her if she would find me a real estate agent who specialized in the north side of Chicago. She was so funny; she said, "Don't you think you should rent for a while to get to know the city?"

I just didn't see it that way. My mind said otherwise.

I said, "Mom, you do this all the time. You help people who relocate from other cities find homes. I need you to do that for me. Please."

The next day she connected me with a realtor who took me to see four one-bedroom condos. While interviewing at this company, I had met Dave, and he said, "Hey, if you need help with anything, please let me know." He was soon getting married in Ann Arbor, Michigan, and we connected on people and places during the interview since I lived near there. He had made suggestions for areas in Lincoln Park that might work for the best train access, where in the city I might enjoy living, and more.

Once I narrowed my selection to one of two properties, I called Dave. I asked him if he would mind looking at the places with me to give me his opinion. He moved his calendar around and went with us and helped me pick my place!

Dave was immensely generous, because if you have ever been married, you know what the week before your wedding looks like. He barely knew me yet gave me time during the busiest week of his life. I bought a tiny one-bedroom condo on the north side of the city with the help of one of the nicest people I have ever had the privilege of knowing.

I began as an account manager, working with existing clients in helping them use our software to advance their business planning efforts. I loved the job. The people working there were all so smart and nice. It was a terrific company with an amazing culture. I learned so much there and made a handful of friends for life.

It was my first travel job, so I also learned how to manage my life with travel. It was like the cruise ship, a family of really smart, hardworking people. Within four years, I was the VP leading a group of five account managers covering the eastern half of the United States. But, wow, was I in for an eye-opening experience!

Initially, I was the leader of the department while still carrying my accounts. I assumed because there were ten account managers that we all worked about the same. But I was not prepared for the differences in how different people work. Over the course of five months, I developed back issues and an eye twitch and was not sleeping well at night. I would wake up in the middle of the night worried about my accounts or a new direct report or whether a specific sale would happen. The stress was real!

What I learned quickly is not everyone actually worked as hard as I did. I realized I hit my goal every year, while others didn't. I came to understand that while some people may work hard, others have to work harder just to stay even with the rest of the pack. I was beginning to see that I was somehow "different." It wouldn't take long before I also knew that it was time to go.

Tip 2: Improv or Stand-up?

While living in Chicago, I enrolled in The Second City school of improv. It was a five-semester series that taught the principles of improv comedy. I had previously done some stand-up comedy, and I thought I should enroll since I was fascinated with Saturday Night Live.

It was so much more difficult than I imagined. In stand-up work, you are looking for one-liners and always strive to throw out a thought or line to be funny in some way. I quickly learned that improv was the opposite. When you throw out one-liners, they mostly fall flat. They don't advance the scene, which was the overall goal.

I made it through two semesters of classes and realized they had taught me to be a better sales leader. Being a sales leader is like improv, not stand-up. You help your sales team along and have success with their success; it is not about being the one with the one-liners! The lessons I learned in the improv classes became part of my tapestry of leadership skills.

What's the Plan?

After seven years with the Sachs Group and then bouncing around for a year plus, I landed a terrific job as the vice president of business development at a community hospital in Chicago.

It was a really great job for me. I don't think that if I had started in the hospital after graduate school and worked for ten years, I would have been able to reach this level in my career. Because I had worked in a hospital prior, I was valuable to the technology company; then after a few years at the technology company, I was more valuable to the hospital. Sometimes going outside first gives you new skills and makes you more of an expert, if you will.

My first day, I found myself in a meeting regarding the hospital's marketing plan, with all eyes on me, as if in slow motion. I had not realized the marketing department was going to be reporting to me, but it was in my purview.

I sensed how I handled that meeting would set the tone for my success in that job. I used all the skills I had learned in my career to date; I was quiet and thoughtful and made no commitments during that meeting. I asked questions I was not exactly sure the group had considered, and people seemed to squirm. The analysis had been not performed for developing specific strategies, and I didn't want to make the team uncomfortable. I took it offline with individual members and smaller groups to work up our overall plan. I kept digging until I had enough information to help develop a marketing plan for the hospital that I would sell internally to ensure we had financial support from my boss and peers before making any tactical decisions in that marketing meeting.

Up to that point, the organization had been making random investments along the way with no real plan. How can you possibly achieve your plan or know if you're on track if you don't know where you are going? It was a good example of how using my skills could benefit the organization.

I had finally come to understand, it is a good practice, no matter who you are and what new job you take, to observe and hold your opinions for those first two weeks. Listen, ask questions, and create relationships. Trust builds slowly, so do this work before you come blowing in with all the answers. I also believe this should be balanced with an awareness of your own underlying wiring. You have to embrace and learn to work with your own style. My style is to gather information up front, analyze it, then make recommendations. At the same time, first impressions matter, and this is a way to manage your brand at a new job.

Hardwired for Success?

It wasn't until years later that I realized making that business plan pitch to my parents at ten was not a typical thought process for a child. As I moved through life, many times I felt "weird." I just wasn't like other kids. I didn't understand that we are all created differently. Everyone is born with certain traits and skills. The environment you find yourself in contributes to what traits of those you are able to develop. But it's very difficult, if impossible, to truly change how your brain is hardwired.

We are all wired differently. And we need to remember that. If we take the time to figure out how someone is wired, it helps us understand how to work with them and accomplish goals with them. Supportive parents and families help foster a person's growth and confidence in entrepreneurship. Emotional support, a risk-taking mindset, and beliefs in autonomy are all things we pick up from our families that can be fostered over time.

Years later, as a business owner, I would come to find myself in interviews with potential new staff, probing into how they are wired—which is the culmination of their brain makeup, upbringing, and life experience. One's wiring determines what kind of worker, team player, and achiever they will be.

It's hard when you're not like the other guy. The one who has a lax work ethic, or the one who takes the credit for something you did, or the one who cheats, and so on. But people notice. Especially when you're the one doing the work and reaching the goals. This reminds me of when my first boss, Bob, wanted to promote me to VP at the Sachs Group.

I said, "Bob, I don't have any formal management training," and I asked to be sent to a formal leadership training program.

He sent me to this fabulous leadership training program that had multiple personality tests, trait tests, and even 360 feedback, which is where your boss, your peers, and the people who work for you provide input about your leadership style. Your colleagues answer survey questions about working with you and, in the case of your boss, how you work for him. This information is combined with your individual testing on work style surveys.

At the end of the three-day training program, we had a personal coaching session. When it was my turn to sit down with the woman who did the profile review, she said—and I'll never forget her words—"I hope you are prepared to be promoted throughout your career. You are the profile of the person who succeeds in management and in entrepreneurship." That was interesting to consider. Did I want to be a leader? What did I see for my future? This was my first leadership role, and it sounded like managing teams may be in my future. Did I want to make that transition for my whole career, from individual contributor to leader?

Tip 3: Success in Interviewing

1. Research the company ahead of time; watch every video and review every section of their website. (Scan websites, e.g., Glassdoor.com, where employees review companies.)

2. Be prepared with answers to common interview questions, such as "How did you handle a tough situation at your last job?"

3. During the interview, do not air your grievances with your last company or boss; they are not relevant to your new job.

4. Prior to the interview, learn about each person you are speaking with and research them online. If appropriate, consider asking them personal questions, such as "How long have you been at the company?"

5. After the interview, send thank-you emails or notes, thanking them for their time. (Snail mail is also good, but the timeliness causes the effort to lose impact, so you may want to do both, a short email and a nice handwritten note.)

Trail Markers:

- *Entrepreneurs probably are genetically wired to take risks, but family support lays the groundwork to develop the confidence and skills required to succeed.*
- *First impressions are real and can be managed. In new situations, quietly observe as you learn about your colleagues and determine whom you can trust.*
- *As you transition from individual contributor to leader, your success will be based on the success of others.*
- *Prepare and practice in order to develop new skills. For example, when interviewing, do your research about the company and people, including preparing a relevant question or two about the job or people you are speaking with.*

Chapter 3

You Can Have It All—
Just Not at the Same Time

I grew up with tremendously supportive parents who told me I
could do anything I wanted to do and be anything I wanted to
be. Getting an education, setting personal goals, and making
good use of your talents were the focus in our house. Personal
relationships were rarely the topic of discussion in our home. I was
gregarious and funny and found it easy to make friends.

While I was not the prettiest kid, the smartest kid, or the most
popular kid, I learned how to build my own group. After all, high
school is made up of all kinds of awkward kids. I loved awkward
people because they are just so much more interesting. I didn't try
to date the captain of the football team or the most popular boy
in school. I realized, *Hey, that is not going to happen*. I learned
to target people like me, a bit quirky and struggling to fit in, and
always found my group pretty easily.

I never had a shortage of boyfriends in high school. When I
became a pompom freshman year (I was a very good dancer),
it became even easier. I was making my way to the "fringe" of

the popular people. But I loved to hang out with people I found interesting, who were just trying to make it through, like I was.

My older brother was the opposite. He didn't date much in high school. He was the "perfect child" at home and a very hard worker in both school and his jobs. He had many friends but was shy and did not date until he was out of high school. He was never in trouble with my parents, a very easy kid to raise, according to them.

While I followed suit, I was interested in dating and there was no protocol that had been established yet. My parents never said much about my boyfriends, except when I invited them over for the holidays. I think it drove my mom crazy. She didn't think a beau should come to Christmas dinner, so I was only allowed to invite them for dessert.

I think it was assumed I would marry at some point, but unlike many girls from the previous generation, there was no pressure for me to find a man and settle down. Looking back, I can see how this lack of attention on love relationships shaped my attitudes toward career and family.

Bring Home the Bacon

My home life was at odds with what I was hearing in the media, which told women they should get married. There was an ad that was popular during my formative youth, in the 1970s, for a product called Enjoli. It was for a perfume that claimed it could help you "have it all" in life.

"I can bring home the bacon, fry it up in a pan, and never, never, never let you forget you're a man!" These ads made it seem like women could do it all—fill the traditional woman's role *and* work. We were being sold a bill of goods. Hogwash!

Contrary to the Enjoli commercial, I found it is pretty difficult to have it all. This societal pressure implied that as a woman you had to focus on your appearance, raise the kids, oh, and now you could go get a job, yet still be responsible for satisfying your man. What? I knew that was wrong!

The reality is that when you get the big job you've dreamed of your entire life, you won't have the time needed to raise a family or, in my case, even the time to go on a date.

I experienced this firsthand in my early thirties. I was running client services for a small, early-stage company. Our largest client was being acquired by our second largest client. Combined, these two organizations accounted for two-thirds of the company's revenue. Because of the way our contract was structured, we stood to lose one-third of our revenue (which amounted to $1 million annually) as the result of this acquisition.

I was working on a proposal outlining why it was necessary to keep the contract numbers from being cut. I was flying out on a Monday morning to Connecticut, with one of the founders of my company, to meet with the client to negotiate this deal. I felt immense pressure.

This was a big deal. I had to convince them to keep spending $2 million and not to assume all the services they needed would be covered in a contract that was roughly $900,000. As part of my preparation, I thought through an approach that showed their return, what they needed from us, and how it increased the plan value for members and employers.

I prepared a packet of presentation materials and had it ready for review by my company's founders an entire week prior to our scheduled meeting. There were four original founders, but only three of them needed to review the proposal. All three of them had families and were running to sporting events for teenagers, so they had various intermittent schedules. I suspected it would be easier to collect written feedback.

I was unable to get them to review it during the week, but not for my lack of trying. I knew it was going to become my urgency over the weekend. I told them all week that I was available to speak live anytime day or night except Friday evening. I reiterated the option for them to send me written comments. Sure enough, they wanted to speak at 5:30 Friday evening. And to top it off, this was made known to me via a meeting invitation at 3:30 on Friday afternoon.

I had a date planned that evening with the man who was to become my husband. I was literally on the phone with the three founders saying, "Look, I can do anytime this weekend, and I could have discussed this anytime this week, but I cannot speak now," as my date was ringing the doorbell!

Thankfully, my boss agreed to speak at 9:00 the next morning. She moved the meeting from 5:30 on Friday evening to Saturday morning for all four of us to meet. When the time arrived, it turned out it was only me and my boss on that call. The other two had done their family thing and blown me off all week long, never weighing in on my presentation. Not only that, but this was the old days, so after they approval the proposal, I still had a half day of printing, collating, and binding to do once we landed on the final presentation format. Needless to say, my entire Saturday was consumed with work. I was learning that my carefully planned timing was dependent on others.

Once Ludi grew as a company, I really tried to protect the weekends of my staff. We used Gmail, which had this great feature where you can schedule what time an email goes out. I tried really hard when I caught up on the weekends to schedule the send to my colleagues for Monday morning, so they didn't feel they had to answer on the weekends.

The good news was, after all the stress, I nailed the proposal for combining those two clients into one. I presented a solid business case for how the current work we were doing should be expanded, thereby growing the revenue from the newly combined client. I had been worried sick about getting that deal done. I knew

28

the gravity of losing one third of the revenue, and the crisis had been avoided!

Child Rearing Has a Real Deadline

I eventually married that lovely man waiting for me to get off the phone with my boss. As we began to think about starting a family, I wondered how I could possibly make it work. At that time I was working in business development in a hospital system. Physicians often only scheduled meetings outside their surgical or clinical hours, which meant early-morning coffee meetings and after-work dinner meetings. Most days I left for those meetings by six a.m. and didn't return home until eight or nine p.m.

I wasn't sure a family was possible with my job requirements, but I was closing in on forty, so it was go or no-go decision time. I had noticed many of my friends slowed their roll a bit when they started a family. Being at home and around for family became more important to them than that vice president job. Like it or not, child birthing has a deadline, and that became abundantly clear to me.

I was thirty-nine when I married my husband. Shortly before the wedding, I had a fall due to a skiing injury I'd sustained fifteen years prior. I was going to need surgery, but it would have to wait until after the wedding. I managed to make it through the wedding with a full leg brace under my dress, and we headed off for our honeymoon in Hawaii. I scheduled arthroscopic knee surgery when we returned from our trip.

Everything seemed to go well until I had some leg pain that I wasn't sure was normal. It wasn't. I was one of the not-so-lucky ones who developed postoperative blood clots, an extremely rare outcome for this type of procedure.

I'm so thankful to my physical therapist. My first trip to physical therapy was nearly two weeks after the surgery (that's how they long they waited to start PT back in those days). I mentioned the pain during my session, and my physical therapist suggested I go right to the hospital.

I was immediately admitted because I was at incredibly high risk for the blood clot to move to my lungs. When that happens, the patient may suffer a pulmonary embolism, which is when a clot moves into your lungs. I was very nervous that evening at the hospital fearing to move at all until I passed the twenty-four-hour threshold. Setting aside how scary it was as a near-death experience, it turned out I had two clotting disorders that would have made having children difficult, to say the least.

I had missed my window. I had forty-year-old eggs. I was trying to recover from knee injury and a blood clot. I had waited so long to begin physical therapy, making recovery a big hill to climb. And now I was facing the reality that if I tried to have a baby, the pregnancy would be high risk. I was raised to chase my dreams with the understanding that I could do whatever I set my mind to. No one had told me I really couldn't have it all, but I wish they had.

As I tried to sort through the grief of realizing I was not going to have kids. I consulted others who were in the process of trying to conceive in their thirties and forties. Why had no one told us this? Our parents encouraged us to get the jobs, pursue the career, but neither of us had felt pressure to marry. Or even encouragement to marry. Is that why we started so late thinking about kids?

Tip 4: Facts about Childbearing

The optimal age for childbearing is twenty to thirty-five. Being pregnant after thirty-five increases the likelihood of certain complications, including premature birth, birth defects, and conception of multiples. The chance of miscarriage, spontaneous abortion, pregnancy complications, and adverse pregnancy outcomes increases, according to the National Women's Health Network. To add insult to injury, pregnancy after thirty-five is often termed a geriatric pregnancy, or as the American College of Obstetricians and Gynecologists puts it, "advanced maternal age."

If you want a family, you need to plan carefully to do so when your body can physically allow this to happen. The deadline is not when you finish graduate school, medical school, or even when you get married. There is a hard reality here that was soft played throughout my life.

I waited and missed the family-growing boat. I thought there were no time limits; I could focus on education, get the career started, then find a man and settle down and have a family. For me, some steps took longer, and the job I found myself in made it really hard to contemplate going through adoption once I understood the cards I had been dealt.

You Can Have It All, Just Not at the Same Time

I think my lesson was, you can have it all, but it is probably not realistic to be doing all these big things at once. And it is okay to choose one thing today as your priority and another tomorrow. This may be an unpopular opinion, because society keeps telling us we can have it all. I just don't think you can have it all at the same time. You can have the big job, but you may have to sacrifice having or spending time with your kids.

The reverse is also true. You can have the kids, but you may have to leave home at six a.m. and be absent until nine p.m. if you also want to do the VP job. I don't think it's the same for men. They can do both because they have often a wife to manage the kids and everything at home.

For the majority of women, things really haven't changed over time either. Men may *help* at home, but women seem to be the master manager of the household. Now, I may get some pushback from men; many husbands and dads do help significantly more than men in the past, but I still think the mama arranges the schedule, buys all the clothing and food, takes care of life, if you will. I've seen some women hire all this out, with full-time nanny help and after-hours day care. That's a solution, but it may mean you're still losing precious time with your kids. It's okay to make choices and cut yourself some slack, however, because again, I don't think it is possible to have it all, all at one time.

It also seems as though people in the working world understand these trade-offs better now as well. When I began my career, it was taboo to have a gap on your resume. In today's world it is totally acceptable, and I believe employers understand and accept these gaps. You may have children or a family member who needed special care, which prevented you from working for a time. Particularly in a society where employment is high and unemployment is low, employers need to be more flexible and creative to get the very best talent.

The Higher You Climb, the Less You Control

Life had made my choice for me, so I focused on my career. I worked hard and climbed the ladder. It may seem counterintuitive, but the higher you climb, the less you control your schedule. When I was working as an account manager early in my career, I arranged my own visits. I could take out my calendar and work around personal commitments to schedule my trips. I could schedule to be in New York ending on a Friday so I could visit friends that weekend.

As I brought on employees to take accounts off my hands, and I moved more into a leadership role, I expected a lighter load and continued control of my schedule. What I got was a different load and less control of my schedule. While I didn't have to make as many client visits, I was thrust into co-traveling on tough or important client visits. The account manager negotiated the dates with the client, not me.

I often received a message saying, "Hey, Gail. We have to be there Wednesday." Well, the truth was we had to be there Wednesday not because that was what the client needed, but because the account manager's daughter was graduating on Thursday or because the account manager's sister was coming into town; it just wasn't a priority to make sure in advance it worked for your boss.

The account managers navigated their own schedules with very little regard as to how it impacted mine. It didn't matter to them that the Wednesday meeting might be following a Tuesday when I was across the country. My calendar became very hectic and much harder to control.

I took that experience to heart. I learned to be much more transparent and open with those higher on the ladder than me. When I needed my supervisor to join me on a trip, I asked what worked best for *their* schedule, and I offered those dates to the client. I think this is part of why I have always been promoted early; I managed up. When you manage up, you figure out what you can do to make your boss's life easier.

Think about it this way: your job only exists because your boss cannot cover all the business on their own; they need you. Now, write this part down: it isn't only all about you. So, if you want to get ahead, think about what can you do to make your boss's job easier.

When I first became a manager, I was astounded at the revolving door of people in my office with problems. "I hate this account." "I have too many accounts." "I don't have enough accounts." "I have xyz disease and sometimes I cannot get out of bed on Tuesdays."

And my personal favorite—I need a raise. Seriously, I have been here ten seconds, and you're asking for a raise? It can be overwhelming as a new manager, sorting through the people stuff. A month into my first leadership position, I went to lunch with the boss who promoted me. He was my favorite boss ever.

"Man, I was so easy to manage," I said.

"Yep, you were," he said. "Why do you think you are the leader now?"

If I brought him a problem, I came with a solution. No one wants to be around someone who is negative all the time. Please take note of this point in your work life, your personal life, and even your married life.

Let me say it again. No one wants to be around someone who is negative all the time. The key to getting promoted is being a self-starting, positive person. Do you have people you avoid at work because they always complain? Do you think people ever avoid you because you are a complainer?

I have to think this is also part of being a successful entrepreneur. Entrepreneurs are by nature pretty positive. You wouldn't start a company if you didn't think you could hit those sales targets, build that product, build a team.

You can apply this same principle to your personal life. I catch myself sometimes when I am leaning too hard on a friend, talking too much about a personal problem I cannot get to the bottom of. I keep spinning. I call them every night for five days straight.

Uh oh. Pump the brakes! You can only tap your friends so long. If you continue to struggle with a particular issue, it is time to get a coach or go to a counselor. Even the *Sex and the City* characters knew this to be true. When Carrie was talking so much about Mr. Big, Miranda, Charlotte, and Samantha begged her to go to a counselor.

I Hit My Head on That Darn Glass Ceiling

My crash into the glass ceiling wasn't obvious to me until I began writing this book. I had been working in a hospital system where I was marginalized by the addition of new male team members who were placed in positions above me. I was passed over for a job I should have gotten, in fact one that I had been doing for years. I led business development for a four-hospital system in Chicago. I had a terrific boss who was based in our Massachusetts market.

Together our offices made up the Boston-Chicago market until 2011 when they split the Chicago market from the Massachusetts market. My Massachusetts boss's approach to business was lean because he had always worked in investor-owned health care, where there is less overhead, more financial rigor, and a model used across the company for the structure. My new market president in Chicago had spent his career in not-for-profit hospital leadership, and I feared he was moving too fast in building out a whole structure that could not be financially supported by these four small hospitals.

The company was heavy on overhead with the addition of all new market people and a fancy office on Wacker Drive in downtown Chicago. Previously, we worked inside the hospitals.

We grabbed an office or shared space or used offices of employees who were out, as needed. We had been scrappy, and this new way of working looked a lot less Motel 6 and a lot more Four Seasons.

I knew the financials of the four hospitals we supported, and they could not carry this overhead. I could see things before they happened; I knew six chess moves away it would all collapse. I had handled the job successfully for five years before the new market president joined the company, but I'm sure you know how that story goes. New regime, new game.

When I met this new market president in advance of his starting in the job, we had an hour or so to talk about the hospitals, the structure, how I could be of assistance, etc. You really have to be the go-to person for the CEO if you are running strategy. I didn't get the sense at our first meeting that we were going to be a good fit.

I told my husband that day, I would need to have a new job in a year because I didn't see my style clicking with this new leadership. The new CEO was the definition of a good old boy. I tried to reserve judgment . . . until he created a senior vice president job, with the same title as mine. They were adding a position above me that sounded a whole lot like my position.

When he told me he'd decided to add this position, I knew immediately that it was not his idea. I believe his boss, the CEO of the overall system in Nashville, was behind it, thinking that in some way I must be missing something. Our system liked big doctor deals. That meant buying a group of twenty-five primary care doctors, picking up an orthopedic surgery group of nine, but these groups did not exist in Chicago.

I knew I wasn't missing anything; Chicago was a different market. The only large groups in the city limits were employed by the three largest health systems. The private doctors were one or two physician groups. It was guerrilla warfare in Chicago, and clearly the brass at my health system thought there was something I wasn't doing.

I asked the CEO if I could interview for the new position. He said yes, of course, and shared that he had already hired a search firm and gave me their number. Boy, things were moving fast. A week or so later, I suited up and headed to this search firm to interview. I quickly decided I did not like how the interview went, and in fact, it seemed like the signs were already pointing me to make a change.

Tip 6: Glass Ceiling

What is the glass ceiling? Are you hitting the glass ceiling? How can you get around it? If you are in a job where you feel you are being held back for any reason, where you don't feel you can be successful, it is time to look for a new job.

If you have tried to work around whatever barrier has been presented, and you are not able to get around it, sometimes it is just time to make a change. Hitting a glass ceiling is usually tied to gender: you have the skills and desire to move up, but you are being passed over because you are female.

So often, people stay in jobs where, through no fault of their own, they are not advancing to the next level though they have the required skills. You can decide to stay and slug it out, fighting your way toward justice, or to leave and pursue your goals elsewhere.

Trail Markers:

- *No one person can have it all at the same time. Big job, kids, taking care of a parent, being a good spouse—all require time and energy. Be honest with what you and your family need from you at any given point in time, and cut yourself some slack. We put unrealistic pressure on ourselves!*
- *There are real trade-offs with having a big career and having children, and childbearing has a timeframe you cannot control.*
- *No one wants to be around negative people. Shift your own thinking, which you can control, to a happy state. Like energy will be attracted to you!*
- *If you have reached a career plateau for any reason, take control and change your job.*

Chapter 4

Done Working for the Dude

Sometimes you need a push to make a change in your life, and other times the path seems so clear. I had been in my position with the four-hospital system in Chicago for about five years but knew it was nearing the time to go since the new leadership team had been in place for six months. I would worry about strategizing my exit once I figured out the exact timing.

I was doing great work for this hospital system. I had been identifying ways for us to align with the physician network, and as a result we were bringing new physicians to our system and engaging those there with new structures.

One of my favorite ways to align with physicians, albeit complex initially, was setting up a scenario in which the hospital and the physicians co-manage the service line. In this structure, independent and employed physicians alike can participate in a contract with the hospital to oversee a business line. Essentially, all parties agree to work toward improving the service line within the hospital. In most of these contracts, the physicians are paid hourly for administrative time, and then there is a pool of money available to be shared if the score card identified and approved in advance is successful at the end of the year.

I loved this particular form of alignment; in fact, I put into place the first agreement like this in the state of Illinois. It was very avant-garde. With this experience, I then helped our other markets—Massachusetts, Arizona, and Texas—forge this kind of alignment as well. This setup always took more meetings to explain everyone's role and to get everyone on board with the plan, but it really allowed the physicians to help direct the clinical care and made a real difference for patients by improving the quality of care.

One evening, as I was driving to one of those informational meetings, my phone rang. I was almost to the hospital, but it was my boss, so I answered. He was calling to deliver some not-so-great news. There was to be another layoff, and I needed to prepare a list of everyone who reported to me, what they did, and who might have capacity to do more, all by nine a.m. the next morning. It was nearly six p.m. already, and I still had this meeting ahead of me.

I hung up the phone and kept driving, but the longer I drove, the more frustrated I became. *I want to put myself on that list*, I thought. While my position was part of the corporate office, most of my direct reports were boots on the ground in the hospitals. It felt wrong to reduce staff there for the second time in a row, and it felt wrong to lay off people who were touching the lives of patients. I felt the layoffs also needed to include the corporate office staff.

I was beginning to realize that I had become persona non grata. I had been cut out of the loop. I was no longer part of key strategy meetings. Even my intern, who had been stolen from me in recent weeks, was sitting in on those decision meetings, which irked me to no end. She had been privy to our fates—not I.

My last boss would not have made this kind of move without my opinion of how the physicians would react and how it would impact our strategies and business objectives. He would have had me in the mix to think through how this would impact people, what he was missing, etc. I wondered now, who was representing the physician's perspective in that group?

As I pulled into the hospital, I forced myself to compartmentalize and put the dilemma out of my mind. After all, I cared about the doctors and hospitals we worked with, and I needed to be my best self for the next few hours of meetings. I got back to my car at eight thirty p.m. to drive the forty-five minutes home.

Suddenly It Is Clear

I replayed everything in my mind as I drove home that night. For months, I had done my boss's job for him and tried to hold my head high, but it didn't matter. I was clearly not wanted there. These people were truly going to drive the system into the ground, and I was powerless to stop them. And somehow I had found myself now on the outside.

What happened to my great job? Who were these people? How did I get to this place? Though I truly enjoyed the content of my work, I was slowly coming to the realization that I couldn't be successful if I stayed—and I needed an exit plan. It was a desperate feeling!

Has that ever happened to you—when you get this funny feeling deep in your gut, but you are not quite ready to acknowledge it? When you are faced with a difficult decision, such as the one I described above, there are so many tough questions you ask yourself. For me they looked like this:

- Do I just stuff down my desperation and push through it?
- Can I keep showing up?
- Will I still be able to do a good job when I feel like there is a separate force working against me?

- Am I taking this too personally?
- Should I just rub some dirt on it, do my job, and not take it all so seriously?

I was the only one in my house earning an income, and my job also covered our benefits. I was trying to figure out quickly how long we could make it financially if neither of us were working. The pressure was high, but I finally reached the breaking point; it was time for me to go. I could not put others' names on the cut list.

I called my husband to let him know I was speeding up my plan to leave, by a lot. I planned to throw myself under the bus the next morning. And that's what I did.

I walked straight into the office with only my name in the cut column for the upcoming layoff. I truly believe the gentleman I worked for genuinely liked and needed me. He was shocked. I got emotional. I shared with him that I loved my job, but it felt like we were playing chess, and they were acting out a checkmate situation that I hadn't seen coming. I loved the physicians and the people in the hospitals, but it was time for me to leave. The fear of staying and losing myself was greater than the fear of leaving.

Tip 7: Decision Challenge

Is there a decision you have been struggling to make in your life? Is it time to change jobs, start a company, or leave a romantic relationship? Try this two-day challenge.

Day 1: Get up in the morning imagining you made decision A. Think about it all day. How do you feel? What is your energy like? Are you happy, or do you have a feeling of dread? Make careful notes.

Day 2: Get up in the morning imagining you made decision B and you are living that life. What are you doing? How do you feel? Are you happy or not? Which decision made you happier? Did one of them in retrospect give you angst or make your belly hurt? Using this technique has helped me arrive at several decisions I needed to make in my life. It works especially well not only for big life decisions but even for financial purchases and smaller choices as well.

Flat-footed

I hadn't been actively trying to cultivate my networks, so I felt like I was starting from scratch. All of my previous jobs had been acquired by some sort of personal connection. You have to be out there to find new opportunities, not necessarily actively looking but actively engaging outside your little bubble. I was in a bubble, for sure. I needed to begin exercising my networking muscle and get back to those connections I'd had prior to my most recent job.

I have never been good at looking for a job when I have a job. It creates a huge conflict for me. You begin to think about what could be with a new job, and it distracts from your current one. Sneaking around and trying to search or take interview calls during work hours is exhausting to me, so that meant trying to search in the evening or on the weekend when the rest of the work world is off. I needed to be laid off so I could freely begin the search for a new job.

Weeks went by, and secret meetings took place in the tower on Wacker nearly every day. Hours of meetings. I tried to do my job, remain focused, and keep a positive view, at least outwardly.

My boss finally told me I was on the list. Another week went by before he called me in to let me know the layoff would be announced the following week. He told me the names of the two people on my team who were getting laid off as well. My last day would be at least two weeks later since there were so many physician relationships to transition.

I honestly think he was initially going to have me lay off the staff in my department also on the list. Nope. No way. I sat there in that meeting and stared right at him and said, "Gosh, that will be hard for you to do all those with the others you have in one day."

There was no way I was laying anyone off before I took my own pink slip. The total layoff ended up being roughly fifty people. Only three people touched the corporate office: myself, Tom Panion, and Catherine Gianaro.

Catherine and I had met early on in my stint at this company. She was one of the few people at the local hospital where I had my office who was nice, and she was a great marketing person to boot! We had become friends over the years; in fact, it was she and her husband who invited me to Iceland, as I mentioned earlier in the book.

The other person was Tom, the marketing manager for the employed physician group, and someone I had been friends with for fifteen years. I had encouraged a colleague to hire Tom into this role; he did the marketing for all the employed physician practices. We had worked together in another company years prior. The marketing department is often a layoff target, so I wasn't surprised they were both on the list.

I stopped by Tom's desk after leaving my boss's office and said, "Hey, what are you doing now?"

"I'm going to hang around and see if he needs anything," he said, pointing to my boss's office. Tom liked to be present each evening when the big brass emerged from their hours of secret meetings.

"No, you are not," I said. "Meet me in thirty at Que Rico," the Mexican restaurant in his neighborhood.

He caught my eye and instantly knew the deal, so he packed his bag and followed me out. I got to my car and called Catherine.

"Hey, can you meet me and Tom at the Mexican place?"

"Oh, I'm just leaving work, but sure," she said.

Tip 8: Networking Is a Lifetime Job

Much like working out, networking is a skill you build over time. LinkedIn is a great professional tool to connect and keep in touch with people. As you build your network, take thirty minutes a week to review your home page and comment on those in your network.

Reach out with individual messages. And occasionally reach out to schedule a coffee or lunch. Your professional network will be the key to all your future jobs, and sometimes those who will be the most helpful are people you barely know!

Margaritas for All

Both Tom and Catherine already knew I was leaving because I had told them in advance, but I didn't want them to be blindsided when they learned the next day that they were also leaving. I decided the best way to deliver the news was just to spit it out.

Catherine approached the table. "So, did it happen? Did you get laid off?"

I said, "I'm laid off. Tom's laid off. You are laid off."

I have always found direct is the best path, but I knew this was abrupt. Tom already suspected, but Catherine was pretty shocked since she had been drafting the communication materials for the layoff itself. She slid down into one of the seats.

"What, how, who? Wait, are you sure?" asked Catherine. "Why are they laying off people from the hospital level? How will my hospital continue with what I have put in place?"

"I guess I'm not surprised," said Tom. "Marketing is often the first place to take cuts."

They both handled it well. Getting laid off is an awful, emotional event, and Tom and Catherine were the type of people you want on your team. I wanted them both to be prepared so they could get all their emotions out in a safe space beforehand.

Talk about being weepy. By this time, I had personally cried enough about leaving the physicians and four hospitals I had worked for so hard for six years, and I'd nearly lost it with each physician conversation I'd had for the weeks approaching this point. So I was done crying.

No one got teary that evening; we quickly switched from the denial stage of grief to the angry stage. I really enjoyed working with these two, and this felt like the end of an era. We had all done our best in our jobs, and we sat around telling stories. *Remember this. Remember that.*

By this time, Catherine and I had both called our husbands, who met us at the restaurant. We had consumed a few too many margaritas, so we both needed rides home that night. Luckily, Tom had walked over from his house, so we all headed out when the pitcher dried up! Overall, it ended up being a nice ending to those jobs. To this day, these two people are still two of my best friends.

Right or Clairvoyant? Now What?

In the end, my read was right. Within the next eighteen months, those two men who ran that market slowly drove the system into such a negative financial state that ultimately the whole market staff was eliminated. Every one of those forty people lost their job or were reassigned.

Those four small hospitals could not support the expense of a big corporate office. I often just don't understand this type of decision and lack of planning. It was apparent to me that as the corporate office was growing, it could not be supported financially, hence the multiple rounds of layoffs.

Plus, this regime only focused on cutting costs. How would they grow the market without me there? Have you ever heard the saying "You cannot cut your way to prosperity"? Well, it is true, and growing the top line revenue is so much more fun than focusing on cutting expenses.

So, what do you do when you've orchestrated your way out of your job? Well, you have to decide what's next, and you'd better do it quickly, especially if you're the one paying the house payment, the grocery bills, and the health benefits for you and your spouse. The safe thing to do is to find another J-O-B., of course, so that's where I started.

I spoke with health care consulting firms, I networked, and I put out a lot of feelers. I had reached a pretty high level in my career, and I had become quite accustomed to a nice salary. The consulting firms I spoke to acted like my salary expectations were crazy high. Looking back, though, it was probably just part of their game.

I received one solid offer that was roughly 70 percent of my base salary with promises of bonuses and commissions that might get me back to my former base salary during a good year. All in, it felt hard to take a pay cut without a clear commission upside that I was used to having in prior roles. Accepting that position felt like the safe choice, since I knew how to do the job, but it was less than what I thought I was worth. In the end, it was easy to pass on that opportunity.

During my search, in the back of my head, I began to think about ways I could create product solutions that would solve the problems that exist between the physician and the hospital.

And when I turned down that offer, I heard that brave entrepreneur voice speak up in my head. *Hey, maybe it's time to take all those ideas floating around back there and start your own thing.*

My desire to work for someone else and try to sell them on this opportunity didn't seem to be working either. The more I thought about it, the more I realized that betting on myself was the better bet. I was done working for the dude.

The Plan Comes into Focus

I began working on the business plan for my own business. I took time to ponder solutions that would make the relationship between the hospital and the doctor better. I kept focusing on the incredible amount of money hospitals invest in aligning with physicians, yet they hardly ever measure the outcomes after they are in place.

I knew there was something to this. With a career spent aligning with physicians and setting up mutually beneficial contractual structures, I saw so many ways the hospital made it unnecessarily hard on physicians. It was also incredibly difficult to assess the return on these investments, because the financial systems did not support this type of analysis, since the data sat in different accounting systems. This was a key relationship and of the utmost importance, so I focused on all the ways using technology could make the process smoother and transparent.

I drove to Wisconsin to bike, journal, and walk along the lake by myself. Even the drive was cathartic. I kept focusing on the idea of improving the relationship between physicians and hospitals. What does that look like to the doctor and to the hospital?

A suite of products was coming together in my mind, and it started with one that solved a very basic problem that was a huge pain point for physicians and hospitals alike. When a physician is hired to be a medical director, they are hired to perform certain duties and are typically paid at a set hourly rate for very specific duties.

The problem? Even as late as 2012, the physician was turning in paper time logs to get paid for these services. I knew software could be a fix for this problem. Physicians could use an app on their phone to record time worked. Smartphones were four years old, and apps were in their infancy in the health care world.

I could design software to feed the appropriate list of duties for the physician to choose, specific to each contract. I could build in messaging, remind them of deadlines, cut them off when the deadline had passed, and have a rich history of the payment data as time went on.

I saw clearly how building the payment data would then build the data so that hospitals could answer the questions, Are we aligned with the right physicians? How much are we spending on our alignment program? Are we focused in the right areas given our strategic plan?

A Product Is Born

No more free texting on paper time logs or trying to remember duties performed weeks prior. I knew this software could solve several pain points. I had been mulling this around for a few weeks. Then one beautiful spring afternoon in Chicago, I went outside and sat at the little red picnic table on my back stoop.

Apparently, I do my best thinking outside. I had a stack of note cards and a goal: sketch out the workflow for the physician and the workflow for the hospital to collect the data needed to properly pay physicians using software. I knocked out thirty note card screenshots in roughly ten minutes. It was divine intervention, for sure.

My husband was a graphic artist, so I took him the note cards and showed him how I wanted the navigation to work on the screen and which buttons I thought the product needed. He essentially took each of my note cards and made a wire frame, which essentially means creating a file for each page.

He designed the look and feel of the buttons for the app, with the words I thought described each path and function. It was all swirling around in my brain, but it was becoming really clear. It made sense to me. He made screenshots and saved the whole file, which by this point was maybe twenty-four pages, as a PDF that I could open on my iPad. This file allowed me to act like I was hitting the button so I could walk through how the software would work.

It was 2012, apps were new, the iPhone was only four years old, and the Android was just behind the iPhone. The entire reason I could build this product was that technology was now scalable with the advent of the app store. It was possible to use commercially available products to support faster development, and I could much more economically build this type of product than before the smartphone technology came along.

So now I had a product idea I could sell, but I had to put the company together. Did I have the skills to do this? What steps would I need to take? How could I learn all I needed to know?

Tip 9: Analysis of the Fast Brain Dump

Has anything like my red picnic table experience ever happened to you? Have you been planning something in the back of your mind for so long, like it has been simmering on the back burner, and suddenly it is ready? You sit down to get your thoughts on paper and they just pour out? For me, I suspect it was a bit of divine intervention and the right push to get me out there doing my own thing. It happened so fast, it was impossible to ignore!

Trail Markers:

- *Take the two-day challenge before making tough decisions. Day one, act as if you've decided A. See how you feel, what your body tells you. Day two, act as if you've decided on B instead. Check in with yourself. How do you feel? What is the right decision?*
- *In order to make any change, the fear of changing needs to be outweighed by the fear of not changing. Sometimes, this takes time.*
- *Networking is part of your job as a human; do it early and often.*
- *Give yourself time to consider new opportunities and directions. We all need quiet brain time to support our decision making. Turn off your phone, take a bike ride, or go for a walk in nature.*
- *If you have an idea for a new venture, be disciplined about what problem you're solving, how it will work, what is special about what you are doing, and then what your business model will be (i.e., how will you make money)!*

Chapter 5

Getting Real, Really Fast

Now I needed to build the company. I'm one of those people who commits. Once I decided it was time to bet on myself, I was all in. I was going to turn those picnic-table sketches into a full-blown SaaS company that would make a real impact for hospital-physician alignment.

In large part, this was the culmination of all my work history. I was in a unique position to be able to merge all the necessary elements from the software world and the hospital world. I had worked for software companies, and I knew how they worked. I had interfaced with many people over my twenty years in my career, and I felt comfortable reaching out to all those people.

I had sold to hospitals, so I had a sense for how SaaS companies priced, serviced, and operated. I also had a hospital business background from working inside the hospital. I knew how decisions were made, why they were made, and how to present the business case to move my product through the hospital.

When you think about starting a business, it's important to consider drawing on your experiences. What unique insights do you have from your past experience? What do you see that is missing?

I often hear about people going into business in a new vertical, meaning a new industry to them, and I think to myself, *Wow, they are brave.* When you start a company, you really have to love what you do because it can be a long few years as you work to get it going. Selecting an industry where you already have knowledge and contacts can help make the process easier.

Sweet Spot

I had twenty years of experience in hospitals working with physicians that would help me grow my new business. It was where I had spent my career, both working inside hospitals and for companies selling something to hospitals to help grow the top line revenue. I was very comfortable with hospital-physician relationships, so for me starting a SaaS business geared toward the health care industry was a logical move.

Most people don't know that physicians are, in reality, independent from the hospital. Many physicians own their own business, while others are part of large physician groups. Either way, the hospital has to pay physicians to do work on behalf of the hospital. Often described as physician-hospital alignment, in all cases, the relationship results in a contract between a physician and the hospital. Unfortunately, more often than not, these contracts lead to frustration and overlapping work on both sides.

At the time, I envisioned a better way of managing the specifics around these contracts by making it easier for the physician to track needed items like time logs and for the hospital to track, manage, and pay the physicians. It would be a new way to approach traditional hospital finance departments, as the current financial systems did not allow hospitals or physicians to look at the detail they needed to manage this relationship. I was sure the way I was approaching this problem was truly a new category of product for the market and would bring insight to the hospitals I worked with. This type of automation had not been built before and was a new solution.

I knew if I could prove my software would save time and money for the hospital and ease the physician's administrative workload, it would be a hit. I got to work putting together the pieces of what I would need. The first big question for me was, Would I need to raise money, or could I get clients rolling fast enough that I could use my nest egg to get the company off the ground? I spoke with anyone who would give me time.

We had a few family friends who either ran businesses or seemed to do quite a bit of investing, so I began there. I quickly realized that if I wanted to raise money, I needed to do a ton of research, and I would need to network with people who were private investors. It felt overwhelming to me to try to do that at the same time I was trying to set up and build the company. If I went that route, fundraising would have to become my full-time job, and I would have to put implementing my business plan on hold. But I also figured out quickly that no one would likely invest until I had the business further along.

I considered doing a round of friends and family fundraising, where you ask those you know for seed money. I gingerly tapped a few people I knew and decided, this too would be a big process that required a ton of time. I had spoken to two family friends who had done some angel investing. One was immediately not interested, and the other asked no fewer than thirty questions.

I ran the numbers and realized I would no doubt have to reach out to hundreds of individuals to end up with a list of twenty who would offer an investment. I just did not know these people readily, so it would take work to compile a list and reach out via cold-calling. I envisioned myself trying to manage twenty separate sales cycles to close each person, and it felt daunting. Not only would it be hard to find prospects; it seemed that keeping them in the loop during and after this process could be arduous.

I also worried I would resent the people who did not want to invest. I buckled down and did some calculations. I estimated the money I had saved would last me two years—two years of paying the development team, running the business, and maybe bringing in one or two additional full-time employees. Decision made. I was off to the races! I was going to bootstrap. I would fund my company myself. I was going to use my life savings and take this huge risk.

It's All in a Name

Suddenly I was the product manager, the marketing leader, the salesperson, the financial planner, the accountant, the inside salesperson, the attorney, and the one pulling all the pieces together. I needed to appear to be up and running. I needed a company structure. Would it be an LLC, an S corp, or a C corp?

My first task was to buy a book discussing the benefits of establishing corporations, either the C or S type, versus an LLC, a limited liability corporation. I spent about a week researching the tax benefits and reading what I could find on the pros and cons of each. Every task had this kind of a lift, needing a full week to understand, then make a decision, then move on to the next task.

I needed a company name, a website, a logo, business cards, and letterhead. The product itself needed a name, and it still needed to be built. I needed a development partner; I needed a proposal format; I needed a contract . . . heavens, I needed to land on how I would price this product. Would I trademark it? Was there any intellectual property to protect? Did I need a lawyer? What kind of insurance did I need?

The questions were flying all around me. It was a crazy time, and I literally learned something new every single day. Every new avenue opened a host of new questions, many of which I didn't fully understand. How do you get a website? Who hosts it? Is there a difference? How much does it cost? How do you get emails set up with your company's website domain? I started with the marketing plan and got down to business. It was invigorating to be figuring out something new every day.

One of my most vivid memories revolves around trying to come up with a company name. Oddly enough, the product name came easily to me: DocTime Log, later rebranded the DocTime Suite. It was self-explanatory. Doctors would log their time in an app that would connect to their contract with the hospital. The app would make it easy for them by serving up duties from their contract. The software would know the start and end date of their agreement, what they were contracted to do, what their pay cycle was, and where the payment should post to on the hospital side. It would route for approvals within the hospital according to who was required to approve on each payment, and it would calculate payments against any rules in the contract.

I had a clear vision for the product and the name, but I had a complete block about what to call the company. I had read enough books on starting a business by that point to know I needed to make sure the URL would be available in the United States, and maybe even internationally, or at least not conflict with any other company's URL elsewhere. The name couldn't be offensive in any language.

I tended to favor two or three syllable words because they are easier to remember. I cut out two sets of the alphabet, and my husband and I created words on the dining room table for a few weeks. We researched them in multiple languages. We picked words we liked—trust, secure, reliable, transparent, alignment. Then we ran each one through language translators to see what those words were in other languages.

Finally, we landed on a word we liked; it was actually a Latin word, *ludi*, pronounced like booty, (loo-di). Ludi meant public games for the entertainment of the Roman people. In my mind, physician alignment was kind of a game, so Ludi seemed to fit! Yahtzee! (Yahtzee was a board game from my childhood; when you won the game, you yelled, "Yahtzee!") We had a name! It seemed to be fairly clear in the United States and would likely pop up pretty high on search engines.

Validating the Product in the Market

The next step in my business plan involved a little grassroots market research. I thought it was a great product, and I knew the hospitals I had worked at in my past certainly needed the product. I needed to make sure the market needed this product. In other words, would someone buy it? Would many customers buy it?

My target was to meet with twenty CEOs before I decided if this product was good enough to build. I was fortunate to have met many of the local hospital CEOs during my time in my former position. I was able to network through them and other people I knew to get to that target.

I called all the CEOs I knew in Chicago proper and asked if they had fifteen minutes to give me their opinion. I told them I was thinking of starting a company and needed their straightforward opinion because they knew the market so well and I trusted their perspective.

While I knew many of these CEOs, I also networked with physicians I knew or anyone from my LinkedIn connections who could introduce me to others. I mined my connections, asked for introductions, then set to work collecting data. I met in person with the CEOs, but I interviewed the physicians on the phone. I was doing the actual product research with physicians, finding their pain points around turning in paper time logs.

I targeted CEOs to see if I could sell the product to their hospitals. During my fifteen-minute meetings with each CEO, I said, Hey, I have this idea to build software that I believe will solve a pain point with your facility. What I'm asking is, at the end of our fifteen minutes, if you would grade the idea with one of three answers:

1. Gail, that is a good idea; I think someday someone might buy that.
2. Gail, that is ridiculous; no one will ever buy that product.
3. Gail, that is such a good idea; I need it right away!

I managed to complete sixteen visits. When I got my second, "That is such a good idea; I need it right away," I kicked into high gear. I worked hard to move those two yeses to a contract all while I was doing everything else behind the scenes: finding a way to build the product, incorporating, setting up all the necessary business aspects, running down the trademark research, trying to figure out how to pay for it all, and figuring out how long I could go running on my own life savings.

Tip 10: Grassroots Marketing Feedback

Once you identify a market need that you think exists, it is critical to go to the market to see if anyone would actually purchase the product. I found it valuable to call on potential clients and ask for their opinion.

People took their feedback very seriously, and this process helped me ascertain if the product would sell. It is much easier to get a meeting when you're asking an executive for their opinion. It takes the pressure off them, because it isn't a sales call; you are simply asking their opinion.

Market research is valuable at every level of the process: at inception of the idea, once the product is formed to collect user feedback, and to glean feedback on how you will price the product. Many companies fail not because the product was a bad idea but because the market fit wasn't clearly outlined for the product.

Fast Track to the Build

With the name, marketing plan, and market validation in process, I needed to get my product built. I called two developers I knew from previous jobs. One was "kind" enough to offer to work for me nights, after his real job, at $125 an hour, which by the way was not at all affordable. At the time you could find development resources at $25–$100 an hour if you went overseas or $50–$200 an hour if you preferred to stay in the United States.

The other developer, and my favorite of the two, thought I was on to something and agreed to give me his time on nights and weekends for a piece of the action. We landed on a number we both thought was fair in terms of a piece of the equity. He was in it for the long haul and would get paid if it worked. We got to business. For six weeks, he was my entire development team.

It was a crazy time. In the midst of all this hustling, I would check in regularly with my development partner. He had a full-time job and was trying to do this at night. It wasn't moving as fast as it needed to be moving. I was at the point with several prospects that I needed to be able to *show* something.

I was starting to get worried that my nights-and-weekend solution for building the product was not going to work. He was a busy guy and didn't have the same urgency I did as I sped through my life savings. He was nearing retirement and was not up for a huge risk of a new job that may or may not work. He did not want to come on board full-time, so I asked him how we could speed this up. He suggested we speak with a development house he trusted that was run by a female developer I had met when I worked with him before.

I met with him and this company in Arizona on August 9, 2012. It was one of the hottest days I had ever experienced: it was 112 degrees, and my blush and face powder inside my luggage turned to liquid in the trunk of my car. We inked the deal, and I stayed for two days as we white-boarded what I wanted to include in this simple solution that was getting more complex by the hour!

It was time to start digging into product development. I wrote more detailed specifications of how I wanted the product to work. For me, I was paying an astounding amount of money to get this thing built, but we needed at least ninety to 120 days to build it, and I was getting closer with both of the initial sales, and new prospects were progressing through the pipeline.

While my partner was working for equity, the development company offered a turnkey solution, and I had to pay them every two weeks. In a "work for hire" fashion, they did everything for me on the IT and infrastructure side. They got my security certificates for the websites. They set up the app in the Apple and Google stores. They set up and maintained the servers. They wrote the code and did bug fixes; they did it all based on my product specifications. It was a "work for hire" scenario, but the intellectual property was mine.

I think I really was fortunate that the leader of our partner company was as conscientious as I was. This company was also a WBE, a woman business entity. The product was built on a solid platform and went on to perform solidly for eight and a half years.

We would later bring development in-house, but we had a very long partnership that worked for me and for the other company. I was, and still am, grateful to my development partner for helping me find a solid group of people who took great pride in their work.

While paying for this every other week was such a huge expense, as it turned out, it was the better deal for me and for the company long-term. They had no stock or ownership in my company. You hear founders say all the time to protect your equity, but that's easier said than done because in the beginning you have no money. The equity only matters if your company is successful years down the road from where you started. But you do only have 100 percent of your company to divvy up. And the time you spend and your personal investment must be taken into consideration.

At the Same Time . . . Still Networking

I created marketing pieces. I set up Salesforce.com for my own outreach and tracking. I tried to secure twenty meetings a month.

The meetings would either be a face-to-face meeting or a webinar with a prospect. I was spending at least six hours a day as the saleslady; then I would change hats and work on the infrastructure pieces needed. But most of my time was spent trying to sell this product. If I could not sell it, I really didn't need to build it, right? I needed to make sure I was on the right path, so above all, sales was the job.

Two months in, I had an awakening about my network and who would be helpful and who would not. I grew up with a mother who worked in real estate, and she got offended if a cousin or friend used another agent without calling her. She would get so mad, but then years later as I bought homes on my own, you really do typically know a handful of realtors in your network of family and friends. It can be tough to select one. I tried to remember it was just business. But I did find myself getting frustrated with people I knew who would not lend a hand. This especially happened with women.

Tip 11: Networking Part Two

Here is how men network.

"Oh, Gail, I remember you said you have software that does something for doctors. I'm not sure if my buddy at hospital A needs it, but here is his number. Tell him I sent you."

My experience is women network based on more detail.

"Sure, I am happy to help you. Tell me exactly what your software does. I will need to see the software. Now what does that button do? What is the return on the investment for the hospital? What would you say to this person at hospital B? Okay, when I understand it fully, I'll ask hospital B if they have this problem that you solve. And if they do, I'll get back with you."

It is such a different approach that I wonder if we women are doing it wrong. I think the opportunity is for women to behave more like men when they network. Don't filter. Don't judge, seek to understand, or interpret.

Just say, "Hey, I know this really stand-up gal who started a company. Not sure if her product is something you need, but it might be worth considering a conversation." You are not guaranteeing or representing my company for me. You are opening doors; that is all. As women, we can take this lesson to help advance our fellow women more quickly and with less effort.

Sales Are the Currency

I also found I needed relationships inside hospitals—people who would help me get introduced to the right people inside their organizations. I mean, I could send emails or call executives in hospitals who I thought needed my product, but would they take my call? There's so much noise coming at these individuals that you can easily get ignored.

In sales, it's a thousand times easier to convert warm leads than cold ones. I began to work through my LinkedIn connections and through my networks. I would reach out to the people I already knew and say, "Hey, do you know anyone at XYZ hospital?" Then I would try to get ten minutes with that person to see if they would help me get to the next right person.

It was like hunting for a needle in a haystack, and it led me to another realization. The people who had led largely entitled lives or who had never worked a service job were the ones who were not willing to make an introduction nor help me get to the next level. Strangely enough, this was also true of the people who had been at the same organization for twenty-five years. They never had to go ask others for business. They had worked hard and been promoted. They likely had not even networked over the past twenty years. They probably never had to look for a job. I learned quickly that the people in business development roles or anyone who had ever sold anything, even retail, were typically much more helpful. Throw a sister a bone!

Everything, absolutely everything, cost money. I was funneling twenty-five to thirty thousand dollars a month into the company account to pay for developers, attorneys, software to support the business, my travel to meet with the development team, and sales meetings. I tried to select places within driving distance—Illinois, Michigan, Ohio, Missouri, Wisconsin, Indiana, Iowa, Nebraska—but I would go anywhere if I could get in front of the right people.

When I could, I stayed with friends if I had to go overnight, but most commonly, I would fly out on the first flight of the day and return on the last. The days were long. When I went to a city, I would try to get at least two appointments.

I was a woman on a mission. I was working literally around the clock, but I was having fun. I really believed in this product, in the new category I was creating, in financial management of physician relationships, and in the relationships between physicians and hospitals.

I have huge respect for physicians. If you think about it, they all care about others. They chose to help others as a career. Physicians are wicked smart. You have to be to make it through medical school. They are competitive by nature and terrific problem solvers. What is diagnosing someone with a condition if not a big problem to be solved?

It Is Easy to Panic

During this period, the money was flying out the door, and I would occasionally have money meltdowns. It is hard to watch your life savings dwindle. While I knew it was a good investment, it was also hard to live off my life savings.

I had a former boss and mentor, Ann Mond Johnson, who had started a company and had a very successful exit several years prior. She was an executive at one of the companies I had worked for earlier in my career. She ultimately left that company, started her own, and several years into that venture called to recruit me away from my hospital job.

She hired me to run client services at her company, including implementation, service, client renewal, and upselling clients into new products. When Ann sold her company in 2006, I was able to share in the spoils because I was a stock option holder. When I had the idea to start my company, I called her and met with her in her living room to get her opinion. She always gave me good advice.

I said, "Ann, how can I do this? I am running out of money."

She said, "When you think you are out, focus only on the next two weeks; can you make it?"

Dig deeper and get creative. She shared her experiences about borrowing her mother's jewelry when she went on sales calls so that she looked successful, and she had an entire assortment of credit cards that she would use, so it was a game of getting new cards and borrowing off those. It really helped to have a veteran share her experience!

Validation!

In the end, I was able to move one of the yeses from my CEO tour to a contract—in a sales cycle that began on July 26, 2012, and ended with a signed contract three months later on October 26, 2012. That's a day I will never forget. It seemed so long in the making, but it had been three short months since I had incorporated and only two months after I had contracted to build the product.

This first client was one of my former hospitals, so I had written or been part of the writing of most of their physician agreements. The only slightly uncomfortable issue was . . . the product wasn't done! I had a signed contract for a product nearly finished, but we needed to get it done pronto!

The executive sponsor was a physician CEO, Dr. Jeffrey Steinberg. He knew firsthand the inconvenience of tracking time as a physician and the compliance risks. He was the first to see the path toward automation. He was also a risk taker; he would identify a problem and go solve it. And if it didn't work, he would move on. Not all executives are like that.

The first client you get is also a vote in your favor; it has to be. They trust you will deliver. Dr. Steinberg believed I could deliver a platform and signed a three-year contract with Ludi with two one-year renewal terms! I couldn't and wouldn't let him down.

I began the implementation process myself for the roughly twenty-five contracts we were going to automate. Implementing a business process requires you to take the workflow of how things work currently and then consider how to streamline it into the software to gain efficiencies.

For example, do we need six executives approving each time log, or should our policy require three executives and clearly outline what each is responsible for? You dig in deep to flow-chart how the process works today, then consider what the automation improves and how it should therefore change that business process.

When you move from paper to automation, some of the manual steps should be eliminated, so the workflow improves and changes. Software implementations fail if you try to automate exactly what is happening today. The reason you are automating is to resolve inefficiencies, so this is an important part of the implementation.

I figured I could get the data inputs ready; then when my development company was further along with the product, we could go live with the doctors and hospital staff. In other words, I would interpret from the contracts what variables needed to be calculated, approve and report on the payments, and upload into the software when it was ready. In hindsight, it was such a blessing that I knew the contracts, the physicians, the managers, and even the risk profile of this system, because it made implementation easier.

I took my implementation questions directly to Dr. Steinberg, and in four one-hour implementation calls, we had what we needed from him to get their contracts implemented. It took me about sixty hours of work to get the contracts in the system, but I was motivated. The product was slated to be released the second or third week in November, so I suggested to Dr. Steinberg that I would come on-site to work directly with the physicians in their regular meetings around Thanksgiving. Luckily, the hospital was only three miles from my house, and his secretary got me the information about all the physician meetings.

I would sit in the back of the conference room with my computer, in each of the physician meetings. I would reset their individual passwords, put the app on their phone for them if it was an Android, and if it was an iPhone, I saved the shortcut on their screen so it looked like an app. (Our Apple store approval was still pending at this time, so I saved with the website URL, and it looked like an app on their main page.) I logged in for them and handed the phone back, then moved on to the next physician while they conducted their meetings.

At the end of each meeting, the leader would give me five minutes to show everyone how to log the time they had just spent in the "medical executive meeting" category, for example, and then they knew how to log other items to then get paid. We were automating this process right in front of them, and it was amazing. Within less than a month, we were live and all forty-five physicians were using the software.

Tip 12: Sales Process

Every sales cycle has a process that is followed to get the sale. Part of the process is identifying who the decision makers are. There is typically an executive sponsor; this is the person at the client site, the highest-level person involved in your sale. They are sticking their neck out and hold the highest level of responsibility.

The economic buyer, or the one with the money, is the one who signs the contract and whose department is going to pay for it. The technical buyer is the one who signs off on the technology. There also may be influencers and users who might use the product long-term; they can usually say no during a sales cycle, but they cannot say yes without the other buyers in sync. Hospital sales processes for software are complex, and there are an average of six to eight buyers per sale.

Managing each sales process is truly like managing a project, and that is after you get people to speak with you! Managing the top of the funnel is more a marketing function. Your prospects may see emails or your company name six times before they make the connection or agree to take a call from you. Creativity counts here, for sure, to turn marketing efforts into leads that will speak with you and then into sales-qualified leads, in other words, those that need your solution and are ready to purchase a solution. It is a complex process to find leads, confirm they are qualified, then make it through the customer-buying journey to arrive at a sale!

Trail Markers:

- *Starting a business required a plethora of decisions, some easy, some hard. Put together a list of all the requirements and needed decisions before starting, and be prepared for the list to grow.*
- *Most importantly, determine how you will finance the new venture . . . and finance your life during this time. My helpful formula is multiply the time you think it will take you by two, and the money you estimate you will need by three.*
- *Do market research with potential buyers and users of your service or product. The business will not succeed if you do not get sales, so do not skip this step!*
- *Revenue risk is real. Do not underestimate the stress you will feel in paying people every two weeks. Defining the sales process early will help support your vision. Without paying clients, you cannot realize your dream.*
- *Finding a mentor who will be there when you need to vent or need advice is invaluable.*

Chapter 6

Stuck in the Elevator

From the moment I locked down that first client, I kicked into high gear to get the product out the door and to have all the needed client support in place. After the physicians were trained on the app, I started to prepare the training materials for those who were going to be approving what the physicians submitted.

Training the hospital users, it turns out, was a heavy lift. Since the software was built to streamline the existing operations, we had to make sure we had the approver trail correct and that each person knew their part in the process. We had built the hospital-facing software to do what was needed, but there were many areas I anticipated wanting to build out further. I knew the success of this whole thing hinged on getting a minimally viable product out the door and then improving it from there.

I am a good planner, or project manager, someone who figures out all the steps, puts them in a chart, and tracks their progress. I was aware of the cadence of releasing a product. In the product build period, I worked to figure out how to trademark the product, how to build the marketing pieces, and how to build out the website and the Ludi brand.

I was running around setting up software that we needed to run Ludi—i.e., QuickBooks to manage my business, Salesforce to track my own activity—figuring out how to manage all the business aspects. Sometimes I just wanted to do the same thing twice. It was pretty exhausting, but also rewarding. It was my own company, and I was making it work, packing the business aspects into about four hours a day and working on sales for another six or so each day so we would have more clients.

I felt like I had stepped into the entrepreneurial elevator, and I thought I was on my way. I was feeling good, but the feeling didn't last long. Every time I turned around, I felt like I was getting stuck in that elevator between floors.

Technology Challenges

There were scary things happening all the time. I learned that conference rooms near the radiology department had bad reception, so the connection would hang and the physician could not log in. I had to show the doctors how to get on the hospital network so they could more easily use the app. They rarely knew their Apple ID, so that usually needed to be reset first.

Sometimes the doctors would think something was wrong with the software, but it was most commonly the connection to the internet. I learned how to troubleshoot phone maintenance—closing apps, restarting the phone, helping the doctor update the current operating system—these things all became routine.

There were, of course, legitimate bugs. Most of the physician meetings were at five or six p.m., after the work day had ended.

On my way home from the evening meetings, I would call the development company, tell them what happened, and the team would troubleshoot it overnight. I would wake up typically to a solution they'd found, and then I would figure out how to close the loop with the physician where the issue had surfaced. For example, I would go to the physician's office later that same week to show them how to proceed in the app. It was not uncommon for a doctor to hand me their phone, go see a patient, and swing back by so I could have them open the app or reset a password to get them up and running.

This was 2012, so not all physicians even had smartphones (the iPhone was released in 2008, so apps were new). Of the initial thirty-two doctors we went live with, only twenty-eight had smartphones. The rest used their iPad or computers.

It was very fun, but also very stressful, and I was forced to learn more about technology so I could be the expert in the room. I was not a developer nor was I an information technology guru. I was a business development executive who had an idea to make a process easier by automating it.

I remember training the first group of physicians on the app. I already knew 90 percent of them because I had worked in their system for five years before I started Ludi. Physicians are often not privy to leadership changes, so often they didn't even know I had left my old job seven months before. My guess is that as I rolled out this software, most of them thought I was still employed by the hospital.

Over the next few years, I occasionally got calls from physicians who remembered my cell number, and I always liked solving their problems. One of my favorites was an issue that outlined a larger problem with these contracts. One day, a physician came to me frustrated and said, "I cannot find the duties I need on this teaching agreement."

"Tell me more," I said.

"Well, when I run out of hours on that contract, I just put them on this one since there is no maximum. But this contract doesn't offer those duties in the drop-down menu."

What?! I thought to myself. This was actually a compliance faux pas! I shared with him the regulatory background of Stark Law, which mandated that the hospital was only allowed to pay him for duties in the contract specified. I tried to scare him a bit because physicians are liable for doing this wrong as well as the hospital.

Sometimes physicians take it better from a disinterested third party like Ludi. We were helping the hospital educate physicians on compliance. I circled back with the hospital administration to let them close the loop with the physician. He shouldn't have been doing that, so it was also a good thing we found this out and put an end to it before it became a compliance issue.

Building for Success

I felt like the elevator was moving. And then it stopped on another floor. As we went through the implementation process with the first client, I focused my time on building out the client services model I wanted to use. This was where I had spent much of my non-hospital-based career. I knew how to delight clients and create a flywheel of good success that breeds the need for more product, which creates a deeper relationship that is hard to break. I went about setting up that model.

I knew it was time to bring in assistance; it could not just be me from here on out. About that time a former colleague contacted me. She wanted to move back to Chicago and heard I was running a company. When we worked together in a previous company, she had helped me onboard new clients, so I knew she would understand my model for client service and would completely take off my plate the day-to-day task of servicing clients.

Tip 13: Client Service

Client service is part of the secret sauce for company success. Many organizations think sales is the top dog, and it is true you are nothing without revenue. Once you get clients, delighting them is the way to faster growth. After running this function for three companies in my career, I have these thoughts in mind:

- Hire people to be client success managers who have experience in the exact functional area your problem solves.

- Do not skimp on these salaries.

- These people are like Trojan Horses; the client invites them in to help them use the product to the best extent. Helping clients use your solution begets more need for your solution; then, more solutions surface that might solve additional pain points, making it a constant growth strategy.

- The knowledge learned is key to feature and function improvements you might consider.

- Spend as much as or more than you do on sales team on your client service team to ensure low customer turnover or churn.

My new colleague was such a versatile team member, she would be able to help me with a variety of things in addition to servicing our one client. She would also be the one who could implement future clients so that I could continue to focus on the growth of the company. This was not going to work if we didn't have sales, so I was still trying to spend five to six hours a day on sales activities.

I quickly learned of another thing I did not understand yet: employment. There are a host of formal processes that need to be set up. First, you need to register with the state tax department. Payroll taxes need to be collected each pay period and submitted to the state. Workers' compensation insurance needs to be purchased. Additionally, calculating these taxes is not easy, so I then needed to set up a formal payroll system. I needed to shop for one, set it up, and get things squared away with the state. I was under the gun because she had joined me, and it was time to get her paid for her first two weeks.

As I evaluated payroll systems, my head began to swim. I was about to make a decision that would have longer-term impact on Ludi. As the founder, I was bootstrapping the company, meaning I was using my own personal money. It wouldn't have made sense for me to pay myself out of my own money, so I had not set up the infrastructure needed to have a real employee. In fact, if I did pay myself, I would just pay taxes on my own money, so it made no sense for me to be part of a payroll system at this point in time.

My summary on this project was that workers' compensation insurance needed to be set up once I had an employee. Payroll tax accounts needed to be set up with the state. Taxes need to be accrued and reported, yet I still wasn't sure exactly how that worked. I could select a vendor to help process everything, including withholding state and federal taxes for each employee and sending them by connecting my bank account. There was this extra step of connecting to the bank account that needed a few days to test the connection. It was not, and could not be, an overnight thing; it would take me weeks to get this set up because I was at the state's mercy. I just didn't see how I could solve this at the same time I was focused on getting the company going.

I literally hyperventilated. I remember my husband came into my office just as the realization of what I had done hit me. I went on and on about all the issues. He hugged me and said, "There, there. You will figure it out." Was that all he could say? He was just telling me I would figure it out. I was forty-three years old, and we had been married for four years, but the money being used to pay for the mortgage payments, grocery bills, electric bills, car payments, *and* this crazy business idea was mine.

I needed a partner, not a halfhearted cheerleader! Good grief, if he would have said, "Hey, I'll get a job to ease some of the burden. I'll apply at Home Depot or wherever I need to," I would have felt better, but he didn't.

He just threw it all right back on me. I literally had gotten so whipped up, I needed a paper bag to breathe. How was I going to do this? I was running just in time with everything. I was in full-on "deal with it when it happens, no time to anticipate all the things in advance" mode.

Okay, I Have to Fix This

After I stopped hyperventilating, I called my trusted advisor to get her advice on how I might proceed. The first thing she said was, "Take a deep breath." It is actually funny how many people told me to breathe over the years; I must speak really quickly or something, but that's a thought for another day.

"It sounds like the role you filled is not yet that of an employee but more of a contractor."

It was true; we were still in the stage of "can we make this thing work?" so it did make more sense that she was a contractor.

Ann said, "Why don't you just 1099 her and make her a contractor as the business builds? Then when you hire the next person, you can consider if you are at the employment stage." She gave me a few tips for employing contractors and explained what is required, how it works, etc.

Hallelujah! This was a solution to kick the employment can a bit further down the road. Take my advice and find yourself a trusted mentor who has started a business and been through the process. These people are invaluable resources as you are starting out, since they have been there. It really can make all the difference to have someone to call.

Crisis Avoided

Thankfully, my colleague agreed, and we mustered on with me hitting the sales side again. Our first client was launching more agreements in our software and adding more physicians and users to the platform. I was thrilled—our first client was growing already.

The evening meeting where the physicians were signing their new agreement would be the perfect place for us to train the new doctors on how to use our software. It would also be a perfect time for me to train my new colleague, or contractor. We could load the app on their phones and follow the same process I had used for the initial launch. The great news was, we were officially in the Apple app store, so we were ready to roll with both iPhone and Android users alike. Before I could set everyone up, however, we ran into a problem.

As it turns out, the hospital had sent me the template agreement ahead of time but not the schedules which outlined payments, so I didn't know the exact dollars and exact physicians being included. On our drive over to the meeting, the exact numbers came in. We were sitting outside the room on the seventh floor waiting to be let in, and I found an error in the math. I had put together many physician-hospital co-management deals when I worked within the hospital system, so I was more than familiar with these contracts.

As I sat there in the hallway going over the numbers, I discovered they were $150,000 over, due to the decrease in the number of physicians participating. This type of co-management deal had one contract only, which all parties signed. The overall contract outlined a higher amount than was needed given the final number of physicians participating. They did not need to go this high on the deal, and the physicians were not expecting it.

In addition, if they went in with the number in that contract, they would not only be overpaying, but the larger issue was that they could potentially be out of compliance with fair market value (FMV). Now, that was a problem. I reviewed the schedule outlining the fair market study they had completed with an outside consultant, and it outlined a cap per physician that would be exceeded if they inked this deal.

When we were finally invited into the room, I caught the CEO's eye and asked for a minute. We stepped outside, and I told him what I had found. I had just saved them $150,000 and a possible huge compliance misstep that could've resulted in millions of dollars in fines. If you are doing the math, we'd call this a ten-times return on investment. They invested $15,000 in their contract with us, and I had already saved them $150,000, and we were barely live with the software!

Sometimes You Just Need to Crawl Out

The CEO asked me to go down to admin while he called his assistant to have her reprint the schedules with the right numbers for the physicians to sign. He stepped back into the meeting, and I went down to pick up everything being printed.

About twenty minutes later, I was riding the elevator back up with the stack of contracts and the elevator stopped midfloor. I can't make this stuff up. If you've ever been in an elevator that stops working, you know the level of panic that sets in almost immediately.

I started sweating and panicking, having seen one too many movies where the elevator drops. Oh my goodness, like it felt with my company, I was stuck in the elevator! As I was feeling about my company, I was punching the buttons, but the car was not moving.

I called for help using the elevator phone and was told maintenance was on the way. I am a bit claustrophobic, so small spaces scare me anyway. I used my cell phone to call my colleague, who was still up on the seventh floor, in case the CEO came back out. She stayed with me on the phone so I didn't panic while I was stuck hanging between floors five and six. It was only five minutes, but it felt more like five hours.

The maintenance guy came with a ladder he placed by his side. I kid you not, he opened the doors by prying them open and held them open. He was on the fifth floor, and I was halfway up to the sixth. He had me crawl out of the elevator and down onto the top of the ladder to get back down to the fifth floor.

I scrambled down in my suit, which included skirt and panty hose, ran up the stairs, handed the CEO the paperwork, and we carried on, training the doctors as if nothing had happened. I'm not going to lie: she had to do most of the talking. I was a mess after my claustrophobic brush with death.

Then the Wind Got Knocked Out of Me

I was killing myself to network, generate leads, and talk to anyone I knew in the industry. I did not have the funds to go to conferences, so I applied to speak. When you speak at conferences, most of the time you get to attend for free. If I was accepted as a speaker, I only had to cover my airfare and accommodations; the conference fees were waived.

I was excited to attend my first speaking engagement just a few months after we had gone live with one of our early clients. It was a physician alignment conference, so I thought that would be the best audience I could hope for. I was going to speak to a couple hundred attendees, and I invited the executive sponsor CEO to speak with me live on stage at the conference. He agreed, and I was thrilled.

There was an attorney who was responsible for most of the physician contracting at the system that his hospital was part of. I had tried to pitch my product to this attorney before we had landed our first hospital deal, and she was less than helpful. I am not entirely sure if she thought our product would unearth something in her contracts, if she thought they just didn't need the solution I had created, or if she was just moving too fast to evaluate something new.

I had run into the same resistance from her chief technology officer as well. I'd tried to reach him as I developed the product, but I was unable to speak live with him. Four months after we signed the first hospital client, a memo was sent out to all hospitals in the system ordering them not to buy any software without the approval of the IT department.

And here's the gut punch—the memo actually referenced my hospital CEO having purchased my product and named my company as well. Perhaps my call to the attorney after I had signed the first contract with my CEO had tipped her off, or maybe the memo came from the CTO directly, or maybe it was part of her frustration with this hospital in general, but regardless, it was going to kill any future business with this system.

The memo explained that the IT department and legal are much better at negotiating deals, and if a software was needed, they would use the collective power of the system to buy the software. I couldn't believe it: the CEO I'd signed with was getting his hand slapped in front of his peers.

My client had agreed to pay $15,500 a year, for three years. Even in 2012, for a system of their size, that was not a significant amount of money by any stretch of the imagination. It was certainly within the power of CEOs to authorize this size of payment without discussion. There was no mention of the incredible compliance risk our solution abated, no mention of physicians' dissatisfaction with doing everything manually, and no mention of the dollars we would save the hospitals. They only talked of the expense and not the potential return on this type of investment.

The system could not cancel the contract with us; it was a valid three-year contract. The CEO had the power to sign a small contract in the scope of the decisions he made daily. But by missing the IT check box, he was being reprimanded, and it killed any future inside that system for me. It was such a low amount in the scheme of things that I was shocked they were calling him out on the purchase. I was left wondering how to move forward with other clients.

Can This Really Be Happening?

So there it was. Literally three days before we were flying to Orlando to speak at this conference, he called me with news of this memo. I was afraid he would cancel and not accompany me to the conference, but after we talked through the pros and cons of continuing with our plan, he still spoke with me at the conference. It was a buzzkill nonetheless.

Before the conference, I was visiting my parents in Florida for the weekend, and the plan was for them to drive me to Orlando for the conference. I was so upset; I barely remember the visit. I doubted everything. Would I face the same roadblocks with my other prospects? Why could I not get business from a system where I had worked for six years and knew a lot of people?

Did I need to change my sales process? Would I be able to make this work? Had I wasted all my life savings? How would I counteract the anger of executives who felt they should have been consulted before our software was selected? Clearly, I was not going to get any additional business from that hospital system. But that was my secret model: get into one hospital, do a good job, and win others.

I took a walk on the beach struggling with my confidence. It was time to put on my big girl pants. I had invested too much to turn back. I was all in. I had a great product, a solid business case, great early customer feedback. Somehow I had just gotten stuck in a political play. I was going to have to rub some dirt in my wounds, dust myself off, and find new clients. I would have to go to people I had not worked with so closely to get this thing going, which saddened me, but *Heck*, I thought, *there are more than five thousand hospitals out there, right?*

Cash Is King

The thing I didn't know then is that it would be seven months between my first and second client and then another seven months to my third. Looking back, I am glad I did not know that is how it would shake out. If I knew how long it would actually take to get the revenue flowing, I'm not sure I would have had the guts to start Ludi.

Translation: I didn't know how expensive this would be to get rolling and how long it would take hospitals to make decisions.

Time equals money. Hospitals were tough. There were a lot of people who could say no and very few who had the guts to say yes. Plus, they were busy doing their normal jobs. Our solution streamlined current processes that were spread across multiple departments. It was difficult to determine who the buyer would be at each hospital, so the sales cycle was tremendously long.

I also began to figure out, no one in a hospital gets fired for *not* making a decision, but if they stick their neck out to say yes and it goes badly, they are risking their job and may in fact get fired. So, part of the selling process was not only evangelizing a new category of product but finding the people who were more likely to make a decision for change.

We were building the business case for hospitals that it was time to start looking at how they were paying physicians and encouraging them to move toward automation. I had to make them realize that the fear of doing nothing was worse than the fear of change. I had to create the need for this product internally and help them navigate their own stakeholders to get to a decision.

There was no one department that was the best to target for the sale. I really had to come at it with a solid story and solid

ROI for leadership, operations, physician services, finance, compliance, and legal teams so they would all buy in for the deal to happen.

Creating a New Category

The fastest path forward was finding the executive who saw the vision, understood the potential, and was not afraid to make a decision and try a new software. I started to call these people unicorns. My experience was, it is kind of hard to find these people, the early adopters.

With any new product, service, or change, there is a bell curve of adopters over time. The early adopters are those who have vision and don't need to "see to believe." Using a bell curve and two standard deviations, this puts 5 to 10 percent of the total target population into this bucket, so finding those willing to make decisions and push their organizations to accept risk with a calculated reward is like finding unicorns!

I had created a solution for a problem many in hospitals did not yet recognize. The current financial systems—accounts payable and payroll—did not have the ability to dig into the analytics of what my product, DocTime, was going to surface.

I could pull analytics they couldn't get to at the time:

- Which physicians are your largest contracts with?
- What service lines are the expensive contracts covering?
- How much are we spending by specialty on each service line?
- How much are we spending for on-call services?
- How does this compare to other hospitals in our system?
- What is our total physician expenditure?
- Are we partnered with the right physicians?

These were the questions I was asking when I was working in development at a high level in a hospital system. But many hospitals were not yet questioning this whole line of expense, just assuming it was as it needed to be. It astounded me that no one seemed to be questioning the expense of physician alignment at this level.

How do you know if you are managing your business correctly if you are not measuring what is going on? I realized in my previous work at hospital systems that these questions were going to become more important as hospitals employed and contracted with more physician partners. That was the more comprehensive end game for Ludi.

The first step was to help hospitals move toward automating this analytics process by starting with contracts where the physician had to turn in time logs in order to get paid. I had to sell the hospital executives on the risks of doing this on paper. Or the benefits of what could be with automation. In each case the story was slightly different based on what would appeal to that hospital and that team. I had to position doing nothing as riskier than investing with us.

I knew what the need was: to sell hospital leadership on why automation was necessary. I created a ROI model, using basic questions I could ask a client and then estimate what their return would be. I could show six different types of benefits, so the art was finding which one appealed in each case. The way I have sold everything in my career has been to illustrate the return on the investment. Step by step I go through how things work today, highlight a problem or risk, quantify the cost of that to the hospital, and add it all up for the return on the investment.

Tip 14: Building a Return on Investment (ROI) Model

I learned this ROI model from a company I had worked for where my job was selling sales training. The idea is that you position your product from the perspective of what the return on the investment will be for the client. You learn how their business revenues and expenses flow and how your product would either increase revenue or decrease expenses. Here is the four-step process:

1. **Describe one pain point your product or service solves.** We have a software to prevent errors in calculating physician payments.

2. **Describe how this is handled today.** Today manual time cards are attached to a check request, which someone fills out, adding the necessary payment. They may or may not pull out the contract to review before making the calculation. Errors are missed because of manual systems, the contracts are complex to interpret, there are many steps in the process, etc. Go through this as many times as you have to in order to exhaust all the challenges in your prospect's current system.

3. **Explain what could be.** How would your product eliminate or minimize one or more of these problems? Our software uses variables in the contract to calculate the payments. No maximums are missed. The software prevents errors because the contract variables are loaded and humans are not actually making the calculations.

4. **In doing so, point out the cost savings.** We can eliminate fifteen minutes of staff accounting time per payment. Keep doing the math. If you have three hundred contracts, at twelve payments a year, we can save $22,000 in accounting time. Drop it to the dollars. Keep digging until you can estimate an expense savings or a revenue growth ROI.

While I felt my ROI story was strong, I was not getting the decisions to move forward. I needed to sign up more clients, faster. The education piece of *"why* you should consider our product" first included a whole education process of illustrating what is today's state and what could be. Here I was again, stuck between the floors. When you create a new category, people don't even know they need your product. It takes an incredible amount of education, and it takes the market time to catch up.

Eighteen months in, I realized sales were slow, the process was taking nine months per client to reach a decision from my initial outreach. I was wearing many hats, but more importantly, I needed help on the sales side. The product side was also taking an incredible amount of time and money. It was time to raise some money to have what is often called "more runway." Time to get the plane off the ground!

Trail Markers:
- *Client service teams are the secret sauce to lowering customer churn, to guaranteeing a successful launch, and to ensuring solid testimonials for future business.*
- *When the company reaches the stage of employing as opposed to contracting for resources, there are multiple steps to take to ensure taxes, insurance, payroll, and all HR steps are followed properly. It will take time and effort in this growth stage.*
- *Political battles inside a prospective client location may arise through no fault of your own or your team's work. These events can be deflating, yet also pivotal, as your resilience will grow and serve as the basis for overcoming future client objections.*
- *Creating a solid return on investment model resulting in hard dollar savings is critical in making sales. All products either have to increase your client's ability to increase revenue or have to help decrease your client's expenses, thereby increasing profitability. Building this model and sharing it clearly during the sales cycle will shorten the sales cycle.*

Chapter 7

Accelerator Program Paved Way to Raising Money

This journey was not going to be as simple as I expected. It was going to take longer than I thought. The good thing about getting stuck in the elevator was all that I learned along the way. I was faced with a problem, I figured it out. This journey forced me to think outside the box.

Failure was not an option. And that meant making sales in order to have revenue to support the business. My goal each month was to have twenty sales calls, face-to-face visits, or webinars where I could be on the phone and share a slide deck and possibly a demo if needed. I was tapping everyone I had ever met to try to find people who could make connections for me and those who might be potential purchasers. I had refined my LinkedIn engine, and I was a networking machine.

In one of those meetings, I met with Michael Sachs, who was the founder of Sachs Group. At the time of our meeting, he was the chairman of Sg2 (Sachs Group 2). When I told him about Ludi, I secretly hoped he would say, "Gosh, Gail! That is a terrific idea. I want to invest." It didn't go that way, of course, but something wonderful did happen in that meeting, and it set me on the right path.

Michael told me about a woman who used to work with him named Nina Nashif. She started a company in Chicago called Healthbox. It was an accelerator program focused exclusively on supporting digital health companies. Accelerators were all the rage at this time in 2012.

An accelerator program is a cohort-based program that provides founders with expert mentorship, education, networking, and sometimes funding to fast-track the development of early-stage companies. They are typically programs you join to start and run your business to get intensive resources over a short period of time. He suggested I should reach out to Nina, tell her what I was doing with Ludi, and get her thoughts.

Here I go on another wild goose chase, I thought to myself, but I agreed to reach out. When I made the call, she was actually pumping gas! Are you supposed to use your cell phone when pumping gas? Something about static potential? Sorry, I digress.

We had a lovely chat, and she said, "Why don't you think about Healthbox for you not as a go- or no-go decision on starting the company but as a channel partner, a means to accessing health systems to buy your software?" She was encouraging me to think of the experience as a path to accessing customers.

That made great sense to me, so I took the leap and applied to Healthbox. In that year, if you were selected, you would be one company in a cohort of seven companies, who were all given offices in the same space as Healthbox.

They would bring in potential investors, potential buyers, and other resources to teach you how to create success in your company. You would trade some equity for a small investment from Healthbox, which was actually directly from the sponsors of the cohort—in our case, one of the sponsors was a national not-for-profit health system and a large regional health system in Indianapolis, Indiana.

A large national health plan was also one of the seven sponsors, and while they were a big name with a big pocketbook, I was less interested in them because I was planning to sell to hospitals. All my contacts at the time were in investor-owned (for profit) health systems, so I jumped at the chance to have access to the largest not-for-profit health system in the country. The capital I received from Healthbox allowed me to cover some of our costs, primarily my service team and the development team, for a few months.

Most importantly, Nina became an important resource and mentor for me. While she was younger than I, she had founded this amazing company, Healthbox. She knew the investor world and understood health care at a detailed level. Nina became a member of the board of Ludi and an important part of our story.

Tip 15: Accelerator Programs

Accelerator programs are a wonderful way to help entrepreneurs get their businesses started and growing more quickly. There are multiple models, but essentially an accelerator program brings together the ecosystem of investors, entrepreneurs, and clients together.

The entrepreneur is supported by business leaders who are experts in their specific areas as well as those who can help the entrepreneur consider how to structure their business for success. This includes raising money to get the business started.

What Happens When You're Late to the Game

I learned a lot about raising money during my time at Healthbox. One of the most surprising things was that we were late to the money-raising game because we already had paying clients.

Most companies raise money right out of the gate as they get started. These companies are described as pre-revenue companies. Before they begin generating revenue, they take on capital in order to grow. Ludi had two paying clients and a robust pipeline due to my rather optimistic nature. So we were officially post-revenue.

The six other companies in our cohort were all in earlier stages of development. They were focused on what their product should look like, whom they would sell to, what their business model should be, and how they were going to pay for building the product. I had come at things slightly differently; I call it bootstrapping. I was nearly eighteen months in and was funding everything from my savings rather than having raised money at the start.

It made for some interesting dynamics in the program. Some of the classes and sessions we attended were about those early stages, so less relevant to me. In some ways, it did confirm the path I was on already, however, so it never felt like wasted time.

Participating in the program, being around others, most of whom were at least a decade or two younger, was invigorating. I was forty-six at the time, and most of the others were in their twenties.

It was an incredible time commitment, and the fact was, because I was spending time at Healthbox, I wasn't driving to twenty prospecting meetings every month. That kind of pace is sometimes needed to uncover new leads that will eventually filter through your sales process. Now, though, I was betting on the promise that more qualified introductions from Healthbox would lead to qualified sales cycles.

People Appreciate What They Pay For

We also scored another checkbox in the unusual path column at Healthbox because we hadn't offered free trials as a pilot program.

That was a conscious decision on my part. I'm a firm believer in the philosophy that people appreciate what they pay for. Offering valuable products or services for free sets up awkward expectations. But it does make it faster to get referenceable clients, so it isn't necessarily a bad idea, just not one I was interested in trying in Ludi's early days.

Have you ever gotten something for free and therefore not taken it as seriously as you would if you had purchased it? I just don't think it is human nature to put in as much effort in fully evaluating a product if you get it for free. You know that old saying, "You get what you pay for."

I was successful in getting our first two clients to sign up with solid three-year contracts with two renewal terms, very common in the SaaS world. I knew if we did things for free, it was likely the client would not "value" that service or product as much as if they had invested money in it.

Speaking of money, I was running out quickly. At eighteen months in, my monthly burn rate was growing, and hospitals were taking so long to make IT decisions. With what I learned at Healthbox, I decided to go for a Series A raise. I was only looking for a million dollars. While a large sum of money, this is not a large sum for building software. You have so much upfront investment as you have to build the whole product before you see any revenue.

My conclusion from Healthbox was that this would be the right approach for Ludi. I was already asking hospitals for business, and I was quickly learning that whatever path I was going to take to raise money was going to involve a sales cycle as well, with any potential investor.

Like any sales process, first you needed a pool of prospects, in this case, financial investors; then you needed to meet with them, pitch your company, and try to cycle into their current investment cadence. I also knew I had to move fast; I had roughly six months of money left before I had to do something that felt reckless, like take out a second mortgage or dip into my 401K money.

Get to the Verb

As part of the Healthbox program, we had an investor day at the beginning of the program and then another one at the end. For these sessions, Healthbox invited in potential investors. We each had three to four minutes to pitch our company, and we would cycle through all seven companies in the thirty-minute program.

Each week the members of our cohort pitched their company to the other members while Healthbox staff sat in and helped us hone our *Shark Tank*-style pitch, meaning sales pitch, and our pitch deck, meaning PowerPoint slides we used to prompt that pitch. This was an extremely helpful process for all, and I watched the other companies grow more succinct with their pitch, hitting their points more quickly and more effectively.

I expect my pitch went down the same path. It's funny how you can get feedback from one person that is immediately contradicted by someone else a minute later. I like to call it feedback whiplash! I had done presentations my whole career, but this one was different. It was personal; Ludi was my idea and my baby.

I had already put several hundreds of thousands of dollars of my own money into it. I was getting anxious about the timing of our success and ready to see things happen more quickly. The biggest takeaway I got from honing my pitch was to get to the verb more quickly. I needed to get to the point quickly in my pitch. Boil it all down to high-level points where you tell your founder story, the problem you solve, how you solve it, the business model, and the opportunity.

I wrote and rewrote my sixty-second elevator pitch until I couldn't think of any other ways to say it. But it was valuable work. It made me better, and I even won a spot to present my elevator pitch in a competition where the winner would receive $10,000.

They literally filmed us in the elevator! The elevator didn't move, but we stood in it for a full sixty seconds—no second takes, just one chance. I didn't win the money, but I learned so much from Healthbox and the resources they brought in. I learned a ton from participating in this program, and it advanced the company immensely.

Introductions

From 2012 to 2014, a new trend was catching on in the hospital world. Hospitals across the country were opening innovation offices. This was especially prominent in investor-owned hospital companies, but even some large not-for-profit systems became wise to the fact that ideas for products came from within their physicians and employees. With their large endowments, hospitals had money to invest in some of these bets, so many created small investment portfolios and began to act like venture capital firms. They did so by forming an innovation office.

New positions like Chief Innovation Officer and Director of Innovation were created to look at companies like ours to see if we would be a fit. These hospitals would then seek to invest in the start-up company, which typically meant investing money as well as offering their hospitals as a pilot site for the product.

Those hospitals that couldn't create their own office wanted in on the game. There were so many hospitals in Chicago considering investing in one of the Healthbox cohorts that it became trendy. Healthbox arranged for the seven of us cohort companies to present over and over. It benefited Healthbox as they recruited future cohort sponsors, and it benefited those of us in the cohort by giving us access to potential buyers of our solutions. These accelerators became cool, and hospitals began to realize their investments could actually make a return.

It was through this process that I met Mark, who changed the course of Ludi. He was the chief compliance officer at what was then a four-hospital system in Chicago, which was owned by one of the Healthbox sponsors. All seven of my cohort members presented on-site at this hospital system to seventy-five hospital executives, and then the system leadership decided which companies, if any, they were interested in trying in a pilot fashion and would ask those for another presentation. Of the seven companies, only two of us were invited in for further review.

Mark told me later that when he saw the software, he thought, "This could really save my bacon."

Cue the mental happy dance! He gave us a chance and championed our software into their facilities. It was our first sale over $100,000, and that deal opened a path to many more hospitals in Chicago, and ultimately that system became one of our largest clients with over 160 hospitals.

Wearing Many Hats

While we were focused on participating at Healthbox, trying to find other clients, I began the sales process of raising money. It seemed to me it was largely smoke and mirrors. I had to sell investors on how our product would solve a problem and how big the market could really be. Investors are biased; we all are really. They are biased by what has worked for them in the past, and they look for companies that look like ones that have succeeded.

I took myself out of the sales role while I was at Healthbox to raise investor money because the program took at least twenty hours of my time every week. Then I needed another thirty hours a week to focus on the raising-capital side, our Series A.

A former colleague and friend I had worked with was in between jobs and had some severance, so she offered to help with sales. She had left her job a few months prior and knew hospitals like I did. She joined us in the office at Healthbox when it worked for her schedule. She had a tough commute, so her hours were sporadic, but, good heavens, I was not paying her a salary, so I was just grateful for the help.

I offered her a 25 percent commission rate. Yes, you read that correctly—no base salary, just a huge commission. She was able to help me manage the two big prospects we met at Healthbox, but she wasn't doing the heavy lift of trying to schedule twenty meetings a month, the lift I had done the first year. I was still working on a big potential Nashville deal, but our pipeline was dwindling quickly. She helped me with the strategy for pricing that deal as we spun and spun around options.

Raising money is hard if you are like 99 percent of people out there who are not connected to people who have money. There are some people who have better access to capital, often called the 1 percenters, who have a different path to raising money. These people are usually those who have family connections and introductions through family networks.

Case in point: check out the path Adam Neumann took when he founded WeWork. He had access to family friends who had millions of dollars. He was the exact pedigree of what founders in the early 2000s were supposed to look like, and he had connections.

Adam was able to raise money at the start and create WeWork with other people's money. His idea, but he got it financed because of the crowds he ran in, where there were many people willing and eager to get a piece of the action. I'm not saying he didn't have to work to position his value; he just had better access to capital.

That wasn't the case for me, nor is it the case for most of the brave souls out there trying to turn their idea into a business. It is a full-time job to get an audience with potential investors. There are websites you can put your information into, to make connections.

Then you follow the usual outreach steps you'd follow with any sales call: contact people, try to get meetings, follow up, rinse, and repeat. Any way you slice it, raising money is a full-time job, no matter who you are.

It's a Man's World

The filters used to evaluate investments often leave out a whole section of the universe. It seems to be more of a man's world here. We had a SaaS-based product; investors understood what that was. I was the one who didn't really understand how to play the game, but I was going to figure it out.

I had no clue how much harder it was for women in the tech space; I still don't know if it is harder, but the numbers just show more male-led companies being funded. I kept slugging away not knowing my path was any different from that of others. I just knew I had to figure it out, and I was running out of money, but I knew this solution was needed.

In my Healthbox class, two of the seven companies were founded by female leaders. That is actually a pretty good number, 30 percent of our cohort had female founders. The year I raised money, less than 2 percent of women-led companies were successful in raising venture capital. You might wonder, has this number changed since I raised capital in 2013?

The answer is, not much, as I write in 2024. To this day, women-led companies command less than 9 percent of the venture capital money awarded. We still have much work to do here. The same applies to private equity, where less than 2 percent goes to women-led companies, both in 2013 and in 2024.

Raising money is largely telling a tight financial story, wherein the product testimonials from your early clients serve to illustrate there is a market. It is about financial metrics above all else, and those with the money dive in on the metrics. I haven't quite figured out what the proper solution is, but women often don't seem to have the financial background to tell the story. I think this is the root problem with women companies being less funded by the financial markets.

Healthbox provided great guidance to us in how to build our stories with metrics. In fact, we were teamed up with an analyst from a sister company that worked on the other side of the equation, making investments. They helped outline how we should present the data, how to determine what we were asking for, and what language we needed to use to be successful.

Healthbox also had investor days, where they brought in investors for us to meet. Early in the program we were all set up in our own "booth," and the investors circulated around. You had plenty of time to hone your three-minute overview! This was where I met more than a handful of institutional investors. I treated this like any other sales cycle.

While I met *new* investors, the best fit I found was with an investor I already knew! The former CEO of the health system I worked for had a family investment office that was an active venture capital firm. During my five months of trying to find a capital partner, I had contacted his family office myself directly. At that time, I could not get beyond his people, the gatekeepers.

Eventually, deflated, I contacted a former colleague who knew the CEO well.

He said, "Gail, just call him directly." I did just that, and we began discussions.

How It All Went Down

After eight months, I had two potential investors on the line through all my networking and outreach. One was a new one uncovered during investor days, and one was the VC firm of my former CEO. The goal was to get each investor to make a formal offer to invest, a term sheet.

This term sheet outlines the majority of the mechanics of the deal. It is highly negotiated by both parties. I had two term sheets in hand, and then it took us another five weeks to finalize a term sheet with one of the investors. Then began the deep dive of the venture capital firm. They dug through everything in what is called a due diligence process.

You know that feeling when you are sitting in an exam room, wearing the paper gown, without your underwear on, and you feel totally exposed? That is the feeling I had during this entire process. It was completely invasive, and I literally did not do much else in terms of running the business. I had two colleagues, one focused on serving the clients and one on growing the prospect pipeline so that I could put time here.

At the end of due diligence, both parties arrive at what is called the definitive agreement, which also outlines when the deal will close, how both parties will behave, and how the money will be wired.

Those two weeks before the deal was scheduled to close were some of the most stressful times of my entire journey. We had most of the definitive deal done, but the lawyers would hang on this issue or that issue. I often felt I was compromising more, and time was no longer in my favor. It is part of the game, and you as the one raising money have more urgency than the firm making the investment. The VC firms, or any institutional capital firm for that matter, can slow the process down to their favor. They know you have so much in this deal you will keep caving here or flexing there or compromising here. I survived, closed the round, but now, how to use it to grow and become profitable?

After many trials and tribulations during those trying nine months of raising money, I was able to close on a Series A million-dollar round financed by this office. I closed our round July 1, 2014, roughly two years after I started Ludi. I gave up equity and some control, but now I had a minority investment partner and the capital to get the growth going faster.

I could hire a sales team. I could invest in marketing and in the product and get the engine moving faster. As we came to the end of our six-month pilot with the big Tennessee company, I was so excited by the results. I planned a trip to Nashville. I had pulled the data out of the software and felt like I had documented a solid return on their investment. I was finally over the big bumps in the road, right? Well, not entirely!

Trail Markers:
- *Look for pathways to providing access to potential clients, investors, and partners. Consider programs that help you learn the market and scale faster, e.g., accelerator programs.*
- *Be prepared to describe the problem your product solves in very short bursts. Thirty-second, sixty-second, and two-minute overviews should be memorized by your team. Often called an elevator speech, how do you describe what pain your product solves, who purchases it, and what the return is? Similar to presentations on Shark Tank, you will have to present your speech thousands of time.*
- *Raising money is a full-time job; be prepared to divide and conquer with your team to make sure the daily business is taken care of while you embark on the funding path.*
- *There are several paths to funding your start-up. From raising money in a friends and family round, to angel investors to a full-blown Series A round, there are multiple options. Each will have pros and cons and should be considered in light of your business and your risk tolerance.*

Chapter 8

Let's Get This Plane Off the Ground!

With a small celebration to memorialize the event with my family and the four people working at Ludi, it was time to get this money working for the business.

In early-stage companies, *runway* is a term that is widely used to illustrate how much time you need to get the company profitable. Said another way, the runway time represents an investment of cash you need to fuel the period to profitability. With the new cash in the bank, Ludi's runway was eighteen months.

I also now had a partner, and I had to formalize setting up a board and also learn how to hold monthly board meetings and all that goes along with using other people's money to run the business. I had to show my partner how I saw the next eighteen months unfolding—what investments I would make, how I would deploy the capital, etc.

I next had to learn how to manage a board. The board of directors acts as an oversight body weighing in on the direction Ludi should take. Because of the structure of my deal, the percent of ownership the venture capital firm was awarded was one board seat. I had a seat as the founder and largest shareholder and also controlled the decision of who would be appointed as the third member of the board.

Managing a board can be a very time-consuming activity for the CEO. Monthly, I now had to report financials formally, provide strategy updates, and answer questions from the board. But it was also a wonderful sounding board, a place to get direction and speak honestly. I was fortunate that my board was only a three-person team, which included me, so as boards go, it was fairly easy to navigate.

The real issue for me was figuring out how to deploy the capital raised in order to grow the business. When you raise money, you are asked how you will spend it. We had also provided an annual forecast during the capital raising process, with both a three-year and a five-year projection, so I had a sense of the best way for Ludi to use the money.

I could see Ludi's areas of need in four buckets: marketing, sales, client service, and product. Fair warning: carrying all the buckets by yourself is hard!

Bucket 1: Marketing

Because we were a start-up, it was vital to get the word out about who we were and what we were doing. We needed to be visible at some of the industry meetings and maybe even sponsor a booth at some of the health care conferences. I became a member of some professional organizations to see if that would lead to any discussions that could ultimately transition to sales opportunities.

Personally, I was a marketing machine; I was *on* all the time. It was expensive to get into the trade shows. First you had to buy the booth, then fly at least one person, if not two, to the event; you had to pay for the hotel rooms for all nights, giveaways for the booth, transportation, food, and possibly even entertain prospects over dinner.

All of that added up fast, and I knew it wasn't possible for us to attend all the conferences we were targeting. It was time for some ingenuity. To get to these events most effectively, from a budget standpoint, I applied to speak at the industry conferences I thought would be of most value. I focused on physician alignment conferences through all the health care associations: the Health Care Compliance Association (HCCA), the American Health Lawyers Association (AHLA), and Becker's Hospital Review and Healthcare Financial Management Association (HFMA).

I stuck to my earlier strategy of making sure I always brought a client to speak with me. If I was accepted as a speaker, I could attend the event free of charge. We evaluated who would be attending each conference. If it seemed like a lot of prospects would be there, we might also invest in a booth. The booth investments all-in started on the low end at around $7,500 and on the top end could cost as much as $20,000. The sales team would then try to arrange four in-person discussions for each conference, in order to justify the expense for our attendance.

I worked really hard to be a voice in the industry and establish myself as an authority. In addition to these conferences, I still pitched at least one article to each of the professional associations annually. It not only made us look more credible, but it was great marketing content to share with prospective clients. Having the investment allowed us to expand our outreach, though we kept the same frugal approach.

Lesson Learned: Individual Contributors and Leaders Are Key

One thing I did not do at this juncture was hire a marketing associate. I managed all the support departments myself, including marketing, finance, HR, and product. Instead of hiring employees for support functions, I chose to invest in employees who were client facing. I felt I could manage all the support roles myself.

Looking back, it might have been the right time to bring in a marketing expert, but it is difficult to find people who can work hard and also lead their department when needed. When you are a small company, you really need individual contributors, but you also need people who are generalists. The best type of colleague I've found is a roll-up-your-sleeves contributor who can become the department leader eventually.

In the marketing world I found myself in, there was a level of specificity I couldn't seem to solve for. Digital marketing, general marketing, writers, people who managed the conference schedule, public relations associates, and lead generation associates—I needed someone who was a generalist, not a specialist, and it was hard to find one person who could do all those tasks who also understood the hospital-physician alignment world. So I found it easier just to muck through all these decisions myself.

Tip 16: When to Hire?

This is a delicate balance for any new company. When do you bring on other employees? I tended to wait until I was so overwhelmed before bringing in a new person, when they were truly needed. The crazy thing was, I was so busy, so it was hard to find time to train the person in their role.

Training people takes time, so find the balance that works best with your risk tolerance.

Organizations that take on money in the early stages have more flexibility in adding staff more quickly to fuel this growth.

In hindsight, Ludi may have grown faster if I had brought on a marketing lead earlier and not tried to do it all myself.

At the same time, I was working hard to develop client quotes and case studies from our existing clients. I created a ton of content that could be repurposed. I worked on things that could be used for an outreach campaign to the different audiences we were pursuing, then written up for a blog, and finally used as a marketing piece on sales calls to show what our return on investment might be.

With my former experience in sales and client services, I knew the kind of materials that were important for prospects and clients, so I just churned it out. I managed our PowerPoint sales deck, the holy grail of how we pitched our story to prospects. This deck was always a work in progress as the market was evolving and we were figuring out exactly why our clients bought our product. It evolved over time for sure. We found it needed a fresh look annually, at a minimum. We added in current market data and recent settlement cases and ensured the deck spoke to the current challenges of our customers at the time.

Bucket 2: Sales

On the sales front, I now had the money to pay base salaries for sales associates. I wanted to hire at least two outside salespeople. Many CEOs say you should hire three at a time because it is likely two of them won't work out. I was getting pressure from a board member to hire multiple salespeople at once.

The challenge was going to be finding individual contributors over managers. We also had to make sure to select salespeople who understood that all aspects of the customer journey were going to be their job. They would have to canvas for leads; they'd have to follow up on all leads that came in through our marketing efforts; and most importantly, they had to be scrappy. They had to be willing to do the hard work. We had two women surface through our search. Both I had known from previous jobs, and we chose the one we thought would be the best fit in our company culture.

I resisted the pressure from my board and slow-rolled the process of bringing on additional sales staff rather than adding them all at once. I'm from Missouri, the Show Me State, so I tend to be more fiscally conservative. Salespeople are expensive, and adding them too quickly could have very well run through my entire investment, especially if I had hired three of them and beefed up our marketing efforts at the same time. But the main reason I slow-rolled this effort was the market readiness was not yet there. The market did not yet seem educated on why our product was needed, and it was taking us time to do this education.

Lesson Learned: Market Readiness Impacts Growth

Much of the sales effort was still education. We needed the salespeople to convince hospitals to solve a problem most still did not know they had. In my gut, I just felt that throwing a bunch of people in the mix so early would not speed up the market readiness issue. I felt we quite frankly had quite a bit of education to do first. My more conservative approach was get one person up and running in three to six months, then add the next.

We needed an individual contributor, someone who could roll up their sleeves and go find leads themselves. We had an early-stage product, so enormous amounts of time and education were part of the sales cycle. We had to first convince hospitals that their current processes were a problem. We knew our product solved three specific pain points, so we had to confirm that those pain points were a problem.

Unfortunately, in the early days, many hospital executives were not concerned with the risks we were pointing out. We had to keep digging, learning from our existing clients, and honing our story to illustrate why our product was needed. In 2013, no one in the hospital system woke up every morning and thought, "Hey, we should automate how we pay our physicians." We were building what is often referred to as a new product category. A product most people didn't yet know they needed. While the pain points were beginning to emerge, we were still on the leading edge of a new solution.

I would say that sales roles are the hardest to fill. Any good salesperson worth their salt will be able to tell you what you want to hear, so it is hard to wade through all the noise and find the one that will be successful. I have learned over the years that if it isn't a fit, you need to deal with it early.

My advice is to hire fast and fire faster on the sales front. Over the years, I had been slow to finally reach the decision that it was time for someone to go. As I look back, I should have moved faster because it was the right thing for the company. Perhaps our growth would have been faster if I had hired more salespeople in those early years. It just was a trade-off for me, the stress that all the salaries brought was greater to me than growing it more organically, though it made growth slower.

Tip 17: Salespeople Who Fit

The best salespeople will wear you out. Truly, the best salespeople are super high maintenance, but it is because they are so good. They might need to talk to you directly frequently. Or walk you through their thinking, seemingly all the time.

They are all worth it as long as they deliver the numbers and fit within your culture. Sales roles are the easiest to analyze from a return perspective. They are being measured on the business they bring in, and it is highly visible.

From a culture standpoint, salespeople can motivate your team in the right direction or be actively disengaged and cause you to have to clean up after them. Finding the right fit is an art!

Bucket 3: Customer Service

Customer service was, I believe, the key to much of the success for two previous companies I had worked for. This was my job for most of my career, leading and building client success. Client services is the secret sauce to building a sustainable, successful company.

Many executives do not share my opinion on this and as a result may not put the right resources in this area. I believe when you do customer service right, the company will grow faster. But it's an expensive investment. The model I favored is one in which each client gets an account manager who is well versed in the client's world and who is responsible for helping the client achieve success. Success is defined by the client; it's not a one-size-fits-all definition.

During the implementation of our solution, we asked our main executive sponsor what success would look like in one year, in two years, etc. We found our secret sauce to be the type of individual we hired. We hired master's-prepared health care executives, all of whom had at least a hospital internship under their belt but also had studied in health administration programs around the country.

The master's-prepared health care executives understood how hospitals operated and therefore how important the physician-hospital relationship was. They understood the financial model of both the hospital and the physician. Our first account manager worked with us on a project basis initially, when we were raising money. I hired him knowing we were closing on our round of funding, but I couldn't commit to pay him long term, beyond the ninety-day project period, if for some reason the funding fell through.

Lesson Learned: Find the Right Guy for the Job

We were so lucky to get that account manager for that project, and thankfully, he was flexible to first be a contractor while we finalized our funding. He had just completed his one-year practicum for his MHA program. I actually met him while I was trying to sell our product to a huge investor-owned system in Virginia, where he was working. I had success with getting that system to commit to a pilot with us because he had confirmed their need when I called on the hospital. He became their account manager for us and worked with the physicians and associates in that market once we closed on our VC money.

Because he knew the hospital system, much like I had known our first client, he was able to get them to engage fully, and it was a successful pilot. He was the right choice for the job at exactly the right time. He was a perfect example of the benefits of hiring a really smart person and giving them a chance to succeed and move up the ladder. He quickly became the one the sales team turned to in the late stages of closing deals to show how we worked with clients.

Client service is all about delighting the client and proving the return on their investment in the product. It naturally led to our clients buying more. I often called account management the Trojan horse of the company. The client invited the account manager in to help them achieve success. By servicing the client thoughtfully, we learned about the challenges our product would solve and those we could solve with a new feature or even a new product.

Account managers identified what changes to the product were needed. Our second product idea came from a client trying to jerry-rig our software. I learned from our team what the client was trying to do with our product and how they were working inside our product around the issue they were trying to solve. When we saw what the client wanted to do, we said, "Hey wait, we can build that and make it better for you. And it will save your staff time, to boot." If you go into client services with that attitude in mind, you'll find yourself in a win-win situation.

Tip 18: Sales Churn or Account Management?

I've never understood the battle between the cable companies and the phone company in obtaining new customers.

For example, take internet service. You sign up for a one-year term with an attractive monthly rate with a cable company. At the end of that year, your bill doubles, so you call the cable company to try to get a better deal. Then you find out the pricing you just saw advertised is for brand-new customers, so your rate remains as it is. You then call the phone company and switch to their internet streaming service for a one-year period.

Same thing: each year, you have to change from the phone company to a cable company, or vice versa, to get an appropriate rate. It is as if they all prioritize new-customer acquisition rather than trying to decrease customer churn. It has to be costly to acquire new clients, yet they spend little time trying to keep existing clients.

I think the secret sauce to growing a business is in servicing your clients, trying to delight them in all ways. This high level of customer service will pay off in the client buying more and even in telling others about your company. One of our best sales opportunities is with a person who uses our product at one hospital, then moves to a hospital that does not have DocTime. My approach has been to spend more on account management, knowing it will always pay off.

Bucket 4: Product Development

The last area—product development—was the one I held on to the tightest even after we raised money. The product itself, how it worked, and the pain points it solved were still evolving, so I felt I was the best person to make decisions on what we did and didn't do to the product. But I was doing many jobs at once and needed help here.

Most SaaS companies have a chief technology officer (CTO) and a product team from the start. I was contracting out the development resources, and the product was essentially my vision for how certain issues could be solved based on my twenty years of industry experience. After raising capital, I knew I could use some help, particularly in communicating my vision to the developers, coordinating with clients, and managing the product road map in a more formal fashion.

The first version of our product was operating very well, but we kept thinking of new features to add to it. As we saw different types of physician contracts, we would think of new ways to better serve the client. We took a chunk of our investment funds and focused on improving the product and building out our second product, which came from our experience in working with our clients.

Much like marketing, product development is a huge and very specialized area. We were working with our external development firm, and that seemed to be going well, so I didn't think it was time to hire a CTO. The person I hired for that role would be a leader, and they would need a whole staff if they were to be effective, so I decided to start with the product management side.

I hired a project manager who could grow into a product manager. At this juncture, we needed clients. We had to close some deals if we were going to make it, so tweaking the product had to fall second to getting revenue in the door. A project manager would help me organize the work we were asking the developers to do, be the person to track progress weekly, and help me manage the communication with the developers. Additionally, I could use this person to help me manage our marketing outreach and related activities.

This role served us well for several years. Because we were backed by venture capital, I felt I now had a partner that could maybe open doors and be a sounding board for ideas. They had clearly seen many of their portfolio companies complete successful exits. A successful exit is when a company starts, grows rapidly, and then is sold so it can reach the next level.

Another great benefit for Ludi with our venture partner was they had services that we could purchase—bookkeeping and finance services, human resources support, and even lead generation. That saved me from having to hire in all those specific areas, and it brought us support as we grew.

Lesson Learned: I Was a One-Armed Wallpaper Hanger

Because my employees were pretty much all sales, client services, and product development, I was wearing a ton of hats. As we grew from the original four team members to nine team members over the next few years, I still ran most of these departments myself; I was the leader and the worker in every department.

I was determined that we were going to be successful; I knew we could do it. Failure was certainly not an option now that I had taken a million-dollar investment. The VC firm believed in me, and for the first time I had someone on my side who wanted me to succeed as much as I wanted to!

I was spread really thin, however, and worried constantly about making payroll during those early years. Every two weeks, people needed to be paid. It is astounding how quickly the weeks go by, and I was very committed to the promise I'd made to myself that I would not take on more than that million-dollar contribution.

Within eighteen months of the raise, and almost four years to the day of starting the company, we were at a positive cash flow! We were officially profitable! The 747 was lifting off the ground, but it was still not easy. Even though on paper we had revenue, hospitals were taking a really long time to pay.

Our contracts said the client had to pay us within thirty days, but many clients would draw it out another thirty days or even try pushing it out further. And so it went that I put on another hat. I became the collection agent as well, sending out each invoice and following payments until the money hit our account. Cash truly is king, as the saying goes!

Many founders and company CEOs talk about this pressure, payroll pressure specifically. It is almost like a rite of passage. If you have ever carried this pressure, you can never forget it. It is like an elephant sitting on your chest. Many companies have a line of credit so they can draw on cash when needed.

Ludi was able to avoid this, but I personally had to make changes as needed. I recall one year, things were so tight, I had to flex my own personal minimal draw, as it was one of the few levers I could pull. I was watching the timing of our accounts payable for one payroll in particular.

Because I did not have a line of credit in place, I was ready to borrow against my house even though I knew deep down, now that we had institutional money, it was no longer an option to keep the company afloat with my own funds. In the past I would put money into the firm as needed, but now there was a formal capitalization table in place, so I would have to contact the attorney if I wanted to add money and figure out how to do it.

About this time, our salesman and project manager came to me with software they wanted to buy that would make them more effective in their jobs. I remember not handling it well; in fact, I think I almost bit their heads off.

"How in the world do you think we will pay for this?" I asked. Our revenue growth that year was nonexistent, and who had the most control over revenue? The sales team. I said, "Hey, we can talk about this when you sell something." In hindsight, I am sure I could have handled it better!

Tip 19: Managing Up

When you approach your boss and begin asking for something, and you see their face tighten up, their eyes get wide, and the flush move up their neck, this may not be a good time. Consider bringing up your request at another time, when you are more likely to have a positive outcome!

Depending on your level in the company, you may not be privy to the pressures going on for your boss at the moment. Manage up by reading the room and maybe asking a question or two first.

For example, imagine if the salesperson had said, "Hey, we really want to change our CRM software. How should we present this to you, and what would the appropriate timing be for such a discussion?" I would have said, "That sounds like a great project to consider once our cash flow is positive. Right now I really need you to focus on closing business so we can make that investment. Realistically, that has to be a next-budget-year discussion."

By taking the time to evaluate it entirely without my knowledge, they had spent time on something we could not afford to buy at that time. We ultimately ended up changing CRM systems at their urging, but it was a year later!

Trail Markers:

- *Have a plan on paper with associated budgets for how you will use the invested money you raise. This will be a pivotal part of the actual raise, but then once you close on your round of financing, it is critical; as time is of the essence.*
- *The investments I chose were in the four key areas for our business, the first being contracted services to help build out the support functions, HR, and finance.*
- *Investing in the sales team and tools for the sales team helped us grow the pipeline and ultimately bring on new customers.*
- *Customer service was important as we grew; account managers helped us increase our footprint within existing customers and keep our churn at zero!*
- *Managing who was working on the product, what our product road map looked like, and how to get the technical team to deliver was a huge investment at this stage, as we were growing rapidly.*
- *Making these investments in time and focus while keeping an eye toward profitability is key to ensuring the investment takes you to the next stage!*

Chapter 9

Personal Strife

Six months after we raised our capital, the sales cycle with a really big fish started heating up. I'd read a lot of start-up books since beginning this adventure. You know what they all said? Don't take on larger clients too early. So, of course, you know what I did? That's right. I tried to reel in a big fish.

The Big Fish

There are a lot of reasons the experts tell you to bring on smaller clients first: your processes may not be ironed out enough to handle a large account; you may not have the staff in place; the product may not yet have been used enough to uncover possible issues; the design may not scale for a large client; or the client may take over your product road map, making the product too unique to that client and not universal for others.

These scaling problems are challenging, and they can be exasperated if you have a huge client hitting too early, before you have had a chance to work through some of them on a smaller scale.

I learned all these lessons and then some straight out of the gate. We met a massive for-profit system during our time in the Healthbox offices and began to get traction. We were able to sign a six-month trial project, and the way I worked the contract was it would roll into a three-year contract at the end of the six-month period unless the client gave notice thirty days before the end of the trial.

We spent several months working with the innovation office at this particular hospital system, following the first project we had done with them. I thought the results were amazing. We had compiled six months of data that I felt was quite compelling.

Our product illustrated where there were potential cost savings and revenue growth and provided insight not seen through other financial systems. Our product also helped the hospital system avoid compliance risk inherent in physician-hospital contracts and related payments. Shock and awe, the results were amazing!

I worked harder on that sale than I had ever worked in my life on any sale. I had the head attorney as our executive sponsor, so we were at the right level. He had introduced me to the chief medical officer and the operations side of the physician enterprise. I had met with the physicians as well as folks from operations, legal, compliance, and finance.

We had six months of data and the opportunity to sit with these head executives for ninety minutes in Nashville. The past year's work was culminating in a system pitch. We were proposing a price per market for this client, which would result in a multimillion-dollar sale.

The evening before the big presentation, I flew to Nashville with my VP of business development. I have to pause here because one of the weirdest things in my career happened that evening over dinner.

Timing Is Everything

The night before the big presentation, my business development VP and I went to dinner. Her role the next day was to support me, but the whole presentation was mine. She had been part of the buildup, the planning, and all the discussions that had occurred up to this point. I had a data book to hand out and review, and I was going to show some of the data live in our software, which either can be impressive or can backfire if there are any technical glitches.

Anyone in sales knows this is an incredibly challenging type of presentation. To prepare, you need to spend some time in advance being quiet, calm, and in your head in order to be on your game the next day. You need to get there early to set up. We went to an early dinner at a nearby restaurant. Over dinner, I could tell the VP was first getting animated, then agitated.

She was focusing on a point in time down the road, when her stock would be worth something, and putting pressure on me to make sure she would get what she expected. At the time, we had six clients and were pitching the big fish. Nothing was for certain, yet she chose this time to needle me about what she needed down the road.

I think this is a great lesson for anyone working for someone else. You cannot operate from a position of tone deafness. It is not all about you all the time. Keep in mind that you are working for someone who, if they founded the company, probably made a ton of sacrifices just to get the thing off the ground.

If you put in the work and make yourself valuable to the team, they will reward you when the time is right. There is a time and place for you to bring up what you need. The night before a huge pitch is not the time or place.

I knew at that moment we had reached a weird place where my VP thought she knew how best to run the company. We had a role reversal at hand. Her behavior that evening made it crystal clear to me what had actually started to become a problem over the past year. That evening as I looked at her talking about how much money she needed when I ultimately sold the company, I knew I was going to have to make a change.

The pitch went very well. I talked them through their data. Then I jumped into the software live. I showed them where I'd gotten the numbers, what I would ask as an executive, and then showed them how I drilled into the data. My goal was to raise awareness about the issues I saw for this large system, how Ludi could address those issues, and what they needed to invest to make it happen. I felt like I had done one of the best presentations of my life. If anyone had shown me this kind of business case when I was on the other side, working inside a hospital system, I would have moved forward immediately.

Toward the end of the meeting, I could sense anxiety from the attendees. It was when we got to the proposal and investment portion of the meeting. It seemed like some people just shut down. I was not sure at the time if they were ready to get out and on to their next meeting or what the issue was. It was an odd ending, after what I thought had been an extremely compelling investment pitch. Well, it was compelling to me, but I began to suspect I had missed the mark.

Wait, I'm Selling a Product, Not the Company

The prospect came back with a counter to my proposal at 25 percent of what I had proposed. Yes, you read that right. They offered me a first-year deal of 25 percent of my ask, and not only that, but it would drop to 12 percent after the first year.

Their other nonnegotiable ask was that they invest money and receive stock in my company. At that point in time, I didn't need any more investors. I needed clients. Their position was interesting to me. They essentially said, "We are so big that we will make you successful with our business alone. Therefore, we deserve a piece of the action."

Sometimes these big systems know their bargaining power and are good negotiators, but this deal was really easy to walk away from. If I accepted their offer, I ran the risk of losing my company and getting swallowed up by their size. I also had spoken with other companies, those that had done business with this system early in the game and those that did business with them later. I also had pretty good estimates of what kind of staff I would need to deliver to such a big system. There was no way to service a deal this large within the confines of their offer. The solution I was selling took service, and I didn't want to underdeliver.

I had worked very hard on the proposal, considering what the implementation would take, how many hours would be needed to service them each year over the term of the contract, and how to achieve customer success. While I didn't land that big fish in 2013, I count it as a dodged-bullet moment. I knew that if we agreed to that deal, we would become a department within this massive company. By not doing the deal, I preserved our independence. But it was still a hard pill to swallow, since we had worked very hard to secure more business from this customer and it was not to be.

We did eventually land a big fish the same size five years later, in 2018, when Ludi was much larger and we had our systems and processes in place. We had signed one market of a large not-for-profit health system in 2014, and each year thereafter we added another market within that system. Often hospital systems are structured into markets, a conglomerate of hospitals covering a specific geography. It took a massive amount of determination and sheer grit. We did whatever it took to get in front of other hospitals within that system because if they were interested, we already had a system contract in place at the system level, so the actual signing of the paperwork was easy if more hospitals in the same system were interested.

Our deal was what is often called a hunting license; we had to go sell the product to each individual market, but if they were interested, paperwork was in place. It was less work for both of us on the legal side, since the terms were already negotiated. It was a matter of adding a schedule to the master agreement, still a contract but one without additional legal and IT review. We just had to convince each market to move forward with our software, and then the pricing and paperwork were ready to go.

Eventually, the great work we were doing in this large system bubbled up to the higher-up leaders—the same ones we had been calling on for years—and we were able to parlay the contract into a full system deal.

We added the remaining six markets, roughly double what our existing Ludi business had been prior. The biggest impact for the company was now we had the money to scale. This was a game-changing moment for Ludi. If we had sold a client this size back in 2013, I think the growing pains would have been really hard. In my opinion, it worked out like it should have!

Doing the Right Thing Can Be Hard

Back to my sales VP. I now needed to deal with this. I needed to work on a severance package and let her go.

This was an awful decision to reach, and unfortunately, it ended badly and was also the end of our friendship. Let my experience be a cautionary tale. When you work with friends or people you have worked with in the past, it can be amazing. But it can also transition into your worst nightmare. Being a CEO is full of these kinds of crossroads, and more often than not, doing what's right for the company is hard.

It was unfortunate that while dealing with this challenge, my home life was becoming more stressful as well.

Tip 20: Having the Wrong Person in the Role

As the founder and CEO, no one inside the company is really your friend. It has taken me more than a decade to learn this, but it has certainly been true for me. Power dynamics shift when you are the CEO, and it creates a different relationship.

As the CEO, it was an adjustment a few years in when I figured out that you cannot share everything with your employees. No one else has the span of control you have and the stress you have in making payroll every two weeks. So while they may be great workers, leaders, etc., they will not be vested in the same way you are, nor can you expect them to be. For them it is a job.

Initially, I treated everyone as friends. I soon realized, however, that the sharing of personal lives will not be two-way with employees, since you are the boss. But it also felt weird to me. I was used to being the fun one, the one always included and looked up to. And you may have to mourn that loss when you are the boss. Once I finally realized I needed my outside relationships to counteract this need, things got easier for me.

I also learned that sometimes having the wrong person was harder than having no one at all. My colleagues in my EO group were a good sounding board. The Entrepreneurial Organization is an international group of CEOs who come together in small groups monthly. Joining this group after moving to Nashville helped me balance the challenges I was facing both at work and at home. The group helped me white-board where I was and come to the conclusion that my colleague had to be let go.

I was learning, but by the time you recognize someone needs to go, they really need to go. The right people leverage you and help you get the company going faster. There are only so many hours in the day for all of us. It is never easy, but if you are going to be the "C" as in CEO, you have to do what is right for the company above all else.

Blew a Gasket

My husband was laid off in 2009 and never quite found his footing after that. It put enormous pressure on our relationship that my partner was not participating economically. We didn't have children or other family commitments, so he had spare time to find his calling. I remained hopeful he would find a job so he could help ease the burden.

In all relationships, I am really not confrontational. And certainly at home, I just want peace. Work is hard enough; I don't want to be walking around at home on eggshells. If something is not quite right, I try to change it. If that doesn't work, I take it and take it, and then I cross an imaginary line and I am done. That is kind of my style: second chance, third chance, fourth chance until boom, I am done.

I felt like I was figuring out the work thing, yet I couldn't solve the at-home issues. It was very frustrating. I was making the company work, and we were growing, year over year, most years at 100 percent, until 2016 hit and things seemed to come to a head.

I had tried everything I could think of to open the lines of communication. I finally came to this realization: you cannot make an adult do something they don't want to do. Nor do adults react well to being told what to do. But marriage is a two-way street. Sometimes you may be fifty-fifty, and other times you may give 80 percent because your partner needs more. It is okay, if it balances out.

I had to end my marriage. I was carrying a company; I couldn't afford to be in a relationship that was drawing down my energy without ever giving me relief. It all became about the business. I loaded up the truck and moved to Beverly. Remember that TV show *The Beverly Hillbillies?* While they moved to California, I hit the road for Nashville.

I loaded a U-Haul with boxes, one bed, a desk, and all my plants and moved to Nashville. I left my ex living in my home with my cats and left to save myself. The plan was for him to stay there with the home intact until it sold, and then we could split things appropriately.

I thought being close to my venture firm would be a good move. Also, Nashville is ground zero for many health care firms and many investors, so I knew it was right for the business and maybe right for me personally.

Key Family and Friends

I had so many friends and family members who supported me through this time. My brother was always there, willing to help in any way, even with financial support if needed. My friends in Chicago were there to help me, emotionally and physically. I am sure they were tired of trying to help me manage through the stress of it all.

I was moving out of a house that was for sale at the moment. My company was based in Chicago, where I had an office with four employees. I needed to logistically get myself moved out of Chicago, but I also needed a place to stay there, as I returned every few weeks to check in with the staff. We were largely virtual, but we did have great shared office space and regularly went into the office.

Karen and Roger, Catherine and David, Stacy and JJ, Carol Ann and Keith, John, Tom, all opened their homes to me as I traveled back and forth for the nine months it took to sell my home. My friends were not only there for me to talk as I went through this very tough time, but they loaned me cars and provided support and a shoulder to lean on. They helped me paint, rake leaves, move out of my house; not to mention, they were all there as sounding boards.

It was not easy to come to this decision, and it was a very hard time with a move between states in the mix. I would not have made it through if not for the team of people in my corner.

Crazy Lady in East Nashville

After moving in the summer of 2016, I would walk the streets of East Nashville for exercise each morning. I was so upset about the personal failure of my marriage, I would cry as I walked. I replayed all the events as I walked. I had waited until I was thirty-nine to get married for the first time, so I did not take this decision lightly. And I was raised Catholic; we were not to divorce.

I could not believe I was now part of that club, the divorced club. I really was sure my marriage would work. Breaking up is always hard, but I was losing an amazing in-law, nieces, a nephew, and even their kids. How would this work?

I needed to grieve. I was finding my way as an entrepreneur, but the failure of my marriage was devastating. It was difficult to reconcile. My parents had been married for fifty-six years at this time, and we were taught in my family to make your marriage work. Divorce was not common in my family.

The same track kept playing in my head as I walked every morning. The good news is it was blazing hot that summer. My tears dried quickly as I hiked up and down the hills in the hottest summer in many years! But I replayed all the events in my head over and over as I walked. *Did I not do something I should have? Did I not give it my all? Was there something else I could have done to save the marriage?* I was living in a city where I had no family and knew very few people. I was forty-nine years old and completely starting over. What was I doing?

I would walk for forty-five minutes in what seemed like a sauna. I thought, *Have I moved to an oven?* I then would shower, slap on a happy face, and get to work. One day, I came home from my walk on a Saturday and decided to vacuum while already sweaty.

I remember the exact moment I had this realization: *It is not only what I had not done in my marriage. Why didn't he fight for me?* We were in it together, so it could not have been all my individual fault. Whatever had happened, it was out of my control, and I needed to let it go.

Tip 21: It's Not Me, It's You

The Bachelor is one of my favorite secret pleasures. Well, I guess not so secret anymore. Here is the difference I have noticed in men and women in this process that has taken place on national TV now for, what, seventeen seasons.

When the men let the ladies go, they get into the limo sobbing, "What is wrong with me?" They think the relationship didn't work because they were deficient.

Now, the men, when they are asked to go, they crawl into the limo and say, "I guess I wasn't her guy, I hope she finds happiness." Or this version if angry, "There was no spark anyway." But for men, it is never something they did or didn't do; they figure it just wasn't a fit. I think they have it right!

The year I moved to Nashville was the first year the company did not grow. The events in my personal life cost me a year of business success.

A company cannot lead itself, even with good people. My lack of focus hurt us. We had burned through our one-million-dollar investment capital, but we started breaking even. But cash was so tight for Ludi, and I was not paying myself as I had been paid in the past. I was taking a draw, and I could live on it, but it was also what I made in 1998 and it was 2016!

No One Panic—The CEO Has a Second Job

I took a night job as an usher at the Ryman Auditorium. For those not from Nashville, it is an amazing historic music venue, known as the mother church of country music. I love country music, and I loved working at the Ryman. It paid slightly above minimum wage, $8.50 an hour!

I took the job because I had moved to a new city and didn't know anyone socially. I now had something to do Friday and Saturday evenings. It was a bonus that I got a few dollars bimonthly to help with the two homes I was supporting.

I probably would have stayed at the Ryman if they would have let me work only weekends. They had promoted a young lady to manager, and they were making decisions that were sometimes very hard to understand. The staff would get so wound up as we listened to the manager's new ideas day after day, with processes changing nearly daily, then changing back when they did not work. Remember, I was an usher. Not the CEO. I even had friends at this job! I was able to show up as an employee; it was just a job. I wasn't there to fix anything, so I was able to watch it all go on and not get pulled into the drama that was growing.

Anyway, one of their mandates was that everyone had to work at least four nights a week. They were having trouble covering shifts, so that was the solution. While the weekends are the hardest to staff, no one wants to work them, and here I was saying, "Hey, I'll work every Friday and Saturday evening, but I do not want to work during the week."

The weekend nights, in fact, were often double shifts, so working a weekend should have been the equivalent of working four shifts. I couldn't work four evenings a week. In eleven months I worked ninety-one shows; it was a blast. Once my Chicago home sold and the pressure eased a bit, I knew I needed to throw all my energy into my company, and that meant being up early each day and staying as late as needed.

I was glad that no one working at my company was concerned at all with me taking a second job. No one noticed and said, "Hey, the CEO has another gig. Maybe she is concerned with making payroll." It was real; I certainly was worried about making payroll. We had run out of our investment, and finances were tight.

But I had a dream, and I was going to pull this off. Looking back, I am proud of my leadership through this time. No one panicked. I'm fortunate that the people who were with me didn't bail. They hung in there with me. They believed in the company. I'd hired really well!

Trail Markers:

- *Getting business from a large client too early can be risky. As your company grows in size, you are able to scale different-size clients. Getting a large client too early can put too much pressure on your existing processes and systems.*
- *Hiring friends can be risky also, to the friendship and to the business. If it doesn't work out, the business takes priority.*
- *Having the wrong person in a role can be worse than having no one in the role.*
- *Personal crisis can add to the stress of running a company, layering on stress to potentially exponential levels. At the end of the day, you have to do everything you can to keep your stress managed, and sometimes that means making personal life changes at inopportune times.*
- *The CEO needs to maintain the focus, to keep the company moving in a positive direction. The company will not lead itself; it needs a strong leader.*

Chapter 10

It's Always About The People

The people you choose for your team will either help you succeed or trip you up every time!

For the first six years, everyone we hired had worked with one of us somewhere along the way. We were running fast, and we wanted to make sure employees were self-driven because we worked virtually. We realized it would be too risky to hire someone whose work ethic was unknown or even someone who had not already proven they could work from home.

Oh Dear, We Are Biased

I was very proud of this part of our story, hiring people we knew. Then I read *How to Be an Antiracist* by Ibram X. Kendi. That book made me realize we needed to search harder for diversity, which meant looking outside of our current circles. My attempt to be efficient and hire people I knew was really a form of bias. We strove to be more deliberate about seeking out diversity as a company from that point forward.

It seems obvious, but if you only hire from the pool of people you know, they are likely similar and have similar experiences. If we respond to a friend of the family who says, "Hey, my nephew needs a summer job," or "My son is graduating with a computer degree," it means we are not providing a chance to others we don't know, which is a form of bias. We began to be aggressive with our posting and networking through various universities to improve our diversity. We began to cast a wider net.

Operating virtually means you are not tied to any one geography, so from a talent acquisition standpoint, you can hire employees who live anywhere. The hardest thing is identifying who will be successful in that environment. It takes a certain kind of person to be successful in a virtual company.

Not everyone can work from home productively. It requires an incredible amount of discipline and self-motivation on the employee side and a great deal of trust on the employer's side. One of our executives joined us after having previous experience as an executive recruiter. She had been running executive searches for hospital positions and therefore knew many executives across the country. She became our secret weapon. She was amazing at interviewing and getting to the bottom of how each prospective team member would assimilate and perform.

I became very focused on making sure each person we hired fit into the organization. When it wasn't a good fit, it could be very disruptive to the culture. One of our early sales associates was technically pretty good at her job. But getting to the sale and how she interacted internally with others left a wake of messes that I had to clean up.

She needed support in doing her demonstrations of the software, but the way she would ask a colleague to help her bristled feathers. The cultural fit factor is key in addition to being able to perform the duties of the job at hand. When the benefit of her sales were being outweighed by the effort needed to keep the rest of the organization going, a change had to be made.

Virtual Culture

A virtual company also requires quite a bit of deliberate communication and carefully curated software to support it. Everything we had was in the cloud. At the time, we paid for maybe twenty different pieces of software, from HubSpot, which was our CRM (customer relationship management), to Box, where all our files are kept in shared folders, to Slack, which we used for internal communication.

Slack provides secure internal messaging, and we had a variety of discussions or channels for each department or type of communication channel Slack contained. We even had a communication channel set up called Random, where we encouraged everyone to share weekend stories and baby photos. We had five Ludi babies in 2020, and we had two on the way in 2021. Sharing our lives in this way was part of our virtual culture.

Our entire logistical structure was built so that if an employee lost a computer, we could stand up a new one for them in one day with all access and nothing lost. We were strict about what can and cannot be on work computers; all client data needed to be housed in Box, not on company computers.

Our company culture evolved as we grew, but I'd like to think we started out with culture in mind. I had great role models from companies I'd been part of in the past with positive cultures. For many in today's world that likely doesn't seem uncommon, but at the time, it was quite the innovative business model. We were a tight-knit group, and many of the employees had worked together before at another great company.

I had taken from my experiences in previous jobs and added a few of my own new ideas to the mix to create what I believed was a great culture. We met in person twice a year, since we were virtual, and those meetings were largely social in nature. The agenda was much more targeted at team building and bonding activities as opposed to hardcore product training or strategic work.

We had an all-company meeting every Monday morning. It was always the same set agenda: watercooler talk, industry and competitor news, round table, and then a special segment. In the early years, each person would share what they were working on, but after a while, with twenty-five team members and growing, we had to modify it a bit so that everyone shared one thing they were proud of from the past week or one thing they were looking forward to in the next week. Eventually, the process evolved even further to only one person sharing by department.

The leaders would write up a report in a Slack channel before the company meeting, so we didn't waste time going over updates the team could quickly read. We also had a book club, where we would read a business book together and discuss it during a segment on our Monday calls. I am a lifelong learner, and to me, reading books is a way to learn and get new ideas. We typically read two to three books together a year, which provided us with a common language as well!

When Things Don't Work Out

Things are never all rainbows and unicorns, though, and as careful as you are when you select people, some of them just don't work out. I found it the most difficult to hire for the sales department. Any salesperson worth their weight in salt can snow you during the interview process.

I don't mean to sound negative, and I don't think all salespeople are deceptive in interviews. It is a fact: salespeople are typically personable, ask a ton of questions, and package their story in a credible fashion. It is really hard to ascertain if they are going to be a good fit or not and, quite frankly, whether they will sell or not. Each company is different, and the ability to sell different types of products can come down to whether or not the salesperson has a passion for the problems you solve.

I couldn't have a chapter on hiring without also including what happens when people resign unexpectedly. My next cautionary tale involves a former colleague whom I hired for sales. He joined us and proceeded to close one of the fastest sales on record. He sold our largest account to date within ninety days of joining us.

He was really good at building his own pipeline and worked hard the first year. But I think he was tired of the hard work and had reached a point in his career where he wanted the next big thing. He came to me and wanted to be a sales leader. At the time, we had two sales executives who were newer in their careers, operating more as inside salespeople.

Inside salespeople are responsible for working the earlier stages of the sales channel. Their job is very focused on taking cold leads and trying to start conversations. Once they connect, they hand the lead to an outside salesperson, who takes the qualified lead and moves it through the customer journey. Sometimes inside salespeople advance and want to move into an outside position which typically has more responsibility and more income potential, but more risk. Others prefer to work the earlier channels and can be very effective once they learn their processes.

This gentleman was working directly with both of our inside salespeople, so we struck a deal. He would be named the VP of sales, continue to be an individual contributor, and mentor the two newer careerists. We shook hands on the deal, he had advanced to a VP position.

The plan would allow me to focus for the next six months on product development. We were working on version two of our base software and a new product that would be released as a new module. Because I was handling product, sales, HR, finance, marketing, and pretty much everything else, I was relieved that he would take sales off my plate.

Tip 22: Recruiting for Virtual Companies

There are people who fit in at virtual companies and those who do not—just like there are people who fit well in early-stage companies and those who fit better in large, established companies.

Working virtually requires self-motivation, focus, and the ability to work in an isolated fashion. There are no water cooler discussions, no lunches with colleagues, so fitting in requires more work than with the in-person model.

When a new virtual employee misses meetings, is late, or runs behind on deadlines, you must research why. It will hurt your company to keep them riding along and not participating as expected. In my experience, when the cultural fit is not there, you have to move quickly to resolve the situation, either by getting them to change or by forcing a change, i.e., letting them go.

In the Lurch

Within sixty days of making him VP, he quit, leaving me knee-deep in product development and back in charge of sales. I had taken my eye off the sales ball, and I soon realized the pipeline was dry.

He was not working as hard as he had the first year, and we were again without a solid pipeline. I felt misled by a friend and was astounded my vision was not his. I am so loyal to people, and I assume the reverse is also true. But that was so naive. For many people, work is a job. For me, it was a mission.

I was a believer in the importance of finding the positives in the midst of the negatives. I soon realized that this situation had created an opportunity to rearrange the deck chairs. So rather than hiring a new salesperson, I brought on our first product manager to help me with the new product and the version two release, and I would again step into the sales and sales leader role.

I went back to managing the two salespeople we had and began combing my LinkedIn nightly, reaching out to people I knew from my past jobs to see if I could reconnect and get them to open a door. I worked like I had in the beginning, trying to arrange twenty discussions each month. We worked with brute force in sales, sending out thousands of emails and making hundreds of cold calls to drum up conversations that might convert to sales. We were successfully rebuilding the pipeline.

Tip 23: Organization Structure

Remember that game Tetris? You had to pack all the blocks and not leave any holes. The more you got filled in, the harder it became, because there were fewer moves. You need room to maneuver. This is how I think about a company's organization chart.

When someone leaves or is asked to leave, do we replace the same position, or is there an opportunity to move things around a bit and pack more boxes? Take these opportunities to reevaluate your strategy, and don't be afraid to try different moves.

Company Team Building

More about our company biannual in-person meetings. These meetings alternated between a fun location for the May meeting and Nashville in the fall. From the early years, I really worked hard to plan these meetings economically. Everything from selecting the flights for employees on Southwest Airlines (our unofficial corporate jet) to planning the activities, the rides from the airport, the agenda, who would present and on what topics, who would be paired with whom for dinner, which restaurants we would visit— every aspect was carefully curated.

In the early days, all the ladies would stay at my home in East Nashville, and I would find a home rental for the men nearby. Our first real trip was to Lake Genova, Wisconsin. Everyone had their own rooms; this was a huge step. We followed that up with a fun trip to Kansas City the next year and Lake Tahoe the following year.

The entire purpose of these meetings was to socialize and team build, so there was very little focus on the agenda and more focus on team-building activities. The agenda included each department having a role, and everyone got to present something. We began integrating volunteer activities into our meetings as well, starting with supporting an underpublicized and poorly funded VA program on Veteran's Day in 2019. Another favorite we supported was a charter school in East Nashville. Supporting charities was an important part of our culture.

The culture we fostered was also one of ownership. I believe people like being on a winning team. They like working for something that is attainable and grand. We worked hard to foster this environment by including growth activities at our company meetings. Our team-building activities included surveys that asked employees to describe their work styles and how others worked, in order to learn more about where each other comes from.

I had consultants come and train us on various topics, such as physician compensation, work styles, and generational differences, in order to advance our knowledge. I invested in a meditation class, followed by a year's subscription for each employee to Headspace, a mental health app, at the start of the COVID-19 pandemic. We had a yoga stretch class at one company meeting that was so well liked I hired the instructor to lead a Ludi yoga class every other Friday at noon.

During the pandemic, I sent gift boxes to all employees for our virtual meetings, since we could not get together in person. The marketing team, my office manager, and I would select eight to ten things with a theme, and opening those boxes on the first day of the meeting was always a highlight! We loaded them up with Ludi logo items, something fun with food and snacks, and one larger gift.

One favorite item was a HALO Bolt, which is a battery that allows you to power two USBs and a regular outlet plug for hours in the event of the loss of power. It is also powerful enough to jump-start your car. In the first year after we gave the HALO Bolts, eight employees put them to use in a pinch. They were a huge hit.

Tip 24: Appreciate Your Team

The success of your business is all about the people. We are all at work for so many hours a day; you have to have fun! Thinking of the themes and gifts for the team amused me. I still love hearing about employees' favorite gifts, from beach towels, to backpacks, to hooded sweatshirts. I loved planning and buying gifts for the team.

My advice for business owners: find ways to delight your employees in small ways that make a difference to the team. The semiannual company meetings and gift boxes were my way to make sure I did small things to appreciate my team.

Friends in Business

While I have so many great stories about what we did well on the people side, there have been some really hard decisions and turning points for the company as well. This is my second cautionary tale about hiring a friend. While I spoke earlier of one friend employee I had to let go, it happened again. I had worked with this amazing lady at two different companies before.

Every few years, she would become discontent and have demands. We were usually able to work through them. But the third time this happened, we reached an impasse. Her communication style was impacting other employees in a negative fashion. Essentially, the whole company began walking on eggshells to pacify her. I felt like I had a dust buster in my hand and my job was to clean up after her every move.

Sometimes I wanted to scream, "Hey, I am still cleaning up everything you stirred up on Monday and Tuesday, so please no more problems; it is only Wednesday!"

Everything I could think of to suggest, to change, to negotiate with her fell on deaf ears. The emotional toll it took on me was huge. I kept trying to keep the peace between her and the other executives.

Finally, I hired a coach to come in and work with the leadership team. It was helpful to have an outside party, someone who did not have all the details, step in with exercises and suggestions for us. Through that work it became clear that she was no longer a fit.

Every relationship is like a bank account. You make deposits. Maybe you help a colleague with a project late one night; that's a deposit. Sometimes you have to make withdrawals as well. Maybe you call that same colleague and ask them to cover your accounts while you are out two days on vacation; that's a withdrawal.

Relationship bank accounts cannot be one-sided. If you only talk about what you need, it runs down your bank account. At some point, making too many withdrawals without any deposits leaves your account overdrawn. Once again, I was sweeping under the rug situations that were making me uncomfortable. I kept making excuses, taking it over and over, and I couldn't see any other way out. Her actions did not support the company agenda, and her account was now overdrawn. It was one of the hardest things I have had to do, but I had to put the company first; we had to part ways.

Sometimes It Is Just Time to Go

Bringing on someone who's a bad cultural fit can have huge impacts on other people on the team. It can be a person who seems like a good fit at first but isn't, or it can be an employee who evolves into a negative Nelly. When the latter happens, you have to move swiftly—which, if I'm being honest, I had trouble with because I kept thinking people would redeem themselves—and with kindness.

I have found a method that works, creating a performance improvement plan (PIP) and having a candid conversation that allows the employee to come to their own realization that it's time to make a change in their work or to exit. My experience has been that when things reach the written level of an improvement plan, it isn't going to end well. The employee is probably not enjoying their job either, and for whatever reason, it isn't a good fit.

Over the years I've gotten really good at helping people make it their idea to leave. I sometimes ask off-the-wall questions people don't expect. We had an employee who was not working out for several reasons, so her manager was telling her she was now going to be on a PIP. I sat in on the tough call to tell her about this plan.

Out of the blue, I said, "I know you love the company, but do you like the content of what we do? Do you like speaking with hospitals about their physician alignment? Don't answer that yet. Think about it over the weekend and call me."

That weekend, she called me to say she really was not that passionate about our subject matter. We agreed her last day would be three weeks out on paper, and she was really done at that moment.

It got easier for me to identify people who were not being successful by thinking of them as taking from the company. I came to think of it this way: *Hey, we are generous. We want everyone to succeed, but if you are intentionally hurting our progress, I cannot have you here.* It made it much easier to have the tough call.

In situations where it didn't work out, it was usually because people got too complacent. Whether it was a salesperson who claimed they had an active pipeline when they weren't really working their leads, or a client service person whom the team had decided to work around, no one was protected. We all had to carry our weight.

I was a servant CEO. I was very loyal—sometimes to a fault. I put my team first as any parent would, and I was protective of each and every one of them. For me, it was exciting to have twenty-five people employed because of my idea. I had seen a gap in how hospitals were aligning with physicians and knew there was a way to design software products to close the gap.

What started as a tiny seed of an idea to solve a problem had blossomed into a successful business that provided a livelihood for twenty-five people. I loved each and every employee, and I valued what they brought to the company. I enjoyed learning about their personal lives and their families. I felt like I was living the American Dream. Well, yes, but it was also, scary as well, because as their boss, I felt responsible for them and their families.

Running the company for me was like being the head of a family. I thought of my team as my kids. And I took their needs and their families' needs seriously because their families were my family too. Knowing that this idea I had supported all of these families was a source of joy, but it was also a tremendous responsibility. I knew if Ludi went out of business, my team would get new jobs in a New York minute, but I would have felt that I had failed them.

I care to hear about everyone's weekend, kids, pets, parents, hobbies, new homes, problems with teenagers, challenges with real estate, but I'm the boss and I've learned that the relationship can't fully flow both ways. Pretty early on, if you have success, you will not be one of the crew anymore.

That has been one of the most difficult lessons for me. One day, before you know it, you'll learn your team has set up a Slack channel to discuss something and you don't have access. They don't invite you in like they did in the beginning when it was all for one and one for all.

That's just the reality of growth and success, and that's when you know you've transitioned from colleague to CEO. I put the company first, before everything else. I became so much happier at work once I realized that the other parts of my life outside my career had to balance me as a person.

As challenging as managing the team could be, I really had no idea what was just around the corner. Managing people is always difficult, but throw in a pandemic and watch how all things get more challenging.

Trail Markers:
- *Successful virtual companies require employees who are highly self-motivated and disciplined.*
- *Recognizing early when employees and the company are no longer aligned provides an opportunity to right the ship. When company goals and personal agendas are not aligned, a change needs to be made.*
- *Working with friends can be rewarding, or can lead to the loss of the friendship if things do not work out.*
- *My favorite saying from a former boss was, "You need to change the employee, or you need to change the employee." Meaning, either the employee's behavior needed to change, or the company would need to change the employee.*
- *People are the secret sauce of your company, but they can also be the cause of problems within the company. Early identification of problems is key to creating change.*

Chapter 11

The Tipping Point

At a company meeting in December of 2016, we did an exercise in three groups called Kill the Company. We literally brainstormed possible ways in which our business could suffer to the extent we went out of business. It was a great exercise in helping to minimize risks as we grew as a company. No one in the room had suggested a pandemic might come along and cripple our hospital customers, putting everything at risk.

Fast-forward, it was February of 2020. We had known that a potential virus with serious implications was coming because of our clients being hospitals. There is a command center at each hospital that prepares for and plans for crisis situations. Their simulated operations for large accidents, earthquakes, chemical spills, etc., are part of regular emergency drills. We began hearing in late 2019 that hospitals were preparing for this COVID-19 virus. By March, we were hearing the COVID-19 virus was on the horizon and most definitely coming our way. Internally, the team dubbed it "The Rona" for the *corona*virus.

Shutdown

My first thought, of course, was for my health and safety and that of my family. But as a business owner, there was this whole new level of concern. I was scared. What would this dilemma mean for my company? Would clients keep paying us? Would I be able to navigate these waters as a leader? Would we survive this?

On March 3, 2020, a brutal tornado blew through Nashville, and in fact, one small block from my home there was sheer devastation. The storm was brutal, taking down every other house, it seemed, across a six-block-wide swath and some fourteen hundred utility poles. The storm left a trail of destruction eighty miles long through the center of Tennessee. Lives were lost. Services would take weeks to repair, and so many lost their homes. Many people flocked to Nashville to lend a hand, likely further spreading this new virus that was on the horizon.

Not even one week later, on March 9, 2020, the world shut down. Fortunately, we were already virtual as a team. We had Zoom, join.me, and Webex, and knew how to use them. We were set up in the cloud for all company functions. Our company could certainly continue to function without missing a beat. I knew our infrastructure was covered; we were okay. Or so I thought.

I soon started noticing that while my staff was accustomed to being virtual, suddenly there was a whole new layer of stress emerging. Spouses were now at home during the day, and many of my team were homeschooling their kids. Spouses were trying to learn how to manage their work from home. Kids were trying to find spaces in the home to set up for success with this new homeschool situation. I saw the stress play out in interesting ways. It was all part of the adjustment to our new norm.

But that also meant that many of our clients were now at home too. Suddenly I could reach the finance teams of large organizations because they were working remotely. This was great, but it was also tricky. One of our main contacts at any hospital was typically the chief medical officer. The physicians were swamped at the start of the pandemic holding it all together. For those contacts working on the front lines during this pandemic, it was not a good time for a sales call. And face it, no one was going to buy software as a service to be implemented with physicians while they were working through a pandemic.

How to Pivot

My biggest fear at the beginning was the money would stop. I was afraid our clients would say they couldn't pay us. And then, too, there was the fact that the sales pipeline had dried up literally overnight. The pending contracts that we had ready to close went away.

I had been helping, with our legal team, negotiate one particular deal for six months, and we had gone from 216 redlines down to three final outstanding items. Redlines are comments from the client's attorneys that represent points of differing views in the contract. We had spent $20,000 in legal fees getting to this point with a large investor-owned hospital system. That sale evaporated and never came back. While we all thought this virus would be gone by the end of the year, we had no idea how long it would drag on.

If our largest client stopped paying us, we would have had an issue. While we had close to 300 hospital clients, we had several of size and one fabulous large-dollar client. They paid us four times a year, and the next payment was due at the end of April 2020. A year prior, we had set up an automated payment process with this client, whereby they would autopay us when the invoice was due, thirty days after the invoice was submitted. This process had been seamless. I held my breath. I figured if we could secure those funds as usual, we would be okay.

On April 30, 2020, the payment hit our bank account as planned. Once I saw that hit and the government stimulus money started flowing to support hospitals, I had more confidence. I started doubling down on infrastructure and focused on things we'd never before had the time to solve or finish completely.

I thought, *Let's use this time to get set up for success on the other side of this pandemic.* I decided to hire a chief technology officer (CTO) to help us decide if we should bring the management of our servers in-house. I also suspected it was time to bring software development fully in-house. We were working with an outside vendor that was not able to deliver as fast as we began to need as we grew. I needed a new team member who was versed in coding, infrastructure, cyber risks, and all things IT.

It didn't stop there. Here we were, two weeks into the pandemic invoicing our largest client, and I found myself working with our team to create a new upsell to offer this client. Due to everyone being at home, I pushed the team to continue to streamline and identify where we could help the client automate more.

It worked! The team pitched automating even more with this large client. We kept pushing to take people out of all the manual steps in making payments after the necessary approvals are made. We worked with their IT team to integrate directly with the client's payroll and accounts payable systems, taking out all the manual work on hundreds of thousands of payments each month.

All of their IT people were now at home where we could reach them more easily. At the client site, the team processing payments agreed this would be a good use of time and funds. And deeper integration, standardization, and automation would save the client money in the end. So we used this "downtime" to go deeper with our largest client and any other clients we could reach.

Discount? Are You Kidding Me?

When the government money began to flow, it seemed the hospitals were being propped up financially. It was less certain for

Ludi, as we were not eligible for the first round of the Paycheck Protection Program (PPP), but the hospitals did have government funds, so I felt pretty certain our clients would keep paying.

I don't know why I grabbed the unknown caller on the phone that day, but I did. The caller, a CFO for a Florida hospital, asked for a reduction in their fees. First, I listened and was very respectful and thanked him for all they were doing. I knew their economic situation was changing and they were trying to pull out all stops.

I said, "I share your concerns. Heck, we are a small company, and I am concerned as you are as to how long this will last." I then asked, "Hey, have we reduced your services in any way?"

When he said no, I just said, "Well, I am glad to hear that we are working hard to use this time to help you do maintenance on the software during this time. In fact, we are working harder. Heck, at least you are getting government COVID subsidies to ease the burden. Ludi is not."

They were one of our midsize clients but a very labor-intensive one to support. We had not cut any service; in fact, we had been using this time to work with clients on cleaning up their data, organizing their sites better, and making sure they were using our latest features. I offered to do a deep dive with the client users to see what we might do.

I said, "Look, I want to work with you, but I also want to be here in a year, two years, etc., so might we agree to stay in touch as things unfold?" We agreed to speak in six months and he moved on to the next vendor on his list.

The second call was from a customer who had been with us for nearly five years. It was time to renew their five-year deal in a few months. When they asked for money back on what they had paid us in 2020, nine months prior, I said, "That is hard to do because you have already received the majority of benefits and service from that payment, so while we cannot do that, let's look at your 2021 investment and make that work for you."

Our internal team met, decided to package more value in for the client, and decided if they gave us another three to five years, we would be willing to extend a discount. We began the negotiation with, "You clearly need a win, and we have a way for you to deliver that to your bottom line." We negotiated a five-year term, in lieu of a three-year term, at a reduced rate that was good for all of us.

Tone-deaf

Even though these maneuvers were successful, we soon realized we weren't going to be successful on the new sales side during the pandemic. Deciding to use our software requires the buy-in of the physicians on the front line. We had a brand-new salesman, who literally began work with us on March 9, 2020. It is so difficult to find good salespeople, so we were excited to have him, and we trained him thoroughly during this time and even had the opportunity to be more strategic with his role.

With his outreach, it seemed possible to reach finance people, since they were often at home as well. It was impossible to speak with clinical leaders, and when he caught one or two on the phone, he got an earful. One of my favorites he got was, "Seriously, are you calling me during a pandemic to sell us software?" Why, yes, sir, I am!

Because it seemed our former sales techniques made us look "tone-deaf," we focused on getting our internal data cleaned up, sales outreach polished off, and systems in place for when we came out the other side. We built out content and tried to expand our relationships by linking in with people. We focused on our case studies with clients, our message in the market and how we approached sales calls. We updated every marketing piece we had during this time.

Another hard call I had to make was to let two of our staff go during this time. Both were salespeople, in fact. When the world first stopped, I suggested to the manager, "Let's wait; even letting someone go seems tone-deaf." We probably kept one on three months longer than we would have under normal circumstances and the other longer than we probably should have.

Sometimes, someone is just not a good fit, and waiting never seems to help the situation right itself. I wanted to rebuild our sales team, as it took us nearly a year to fully train each executive, but I wasn't that much of a risk taker to replace the two we had lost. We held with the one who began at the start of the pandemic and let him drive the business where he could, but I did hold on hiring any additional team members for sales until late 2021.

Tip 25: Building the Sales Team

Many organizations have the guts to hire two to three salespeople for every sales role they need filled, knowing that one or two likely won't work out. Sales is critical for your success, because revenue is the only way to grow. Because of my more conservative nature, I only tried this once, in 2018. I hired a sales manager and she hired two outside associates, and we had one more in-house already.

Here is my cautionary tale. With a nine-month average sales cycle, it is difficult to tell if a salesperson is having success until close to the year mark. I let my new sales manager "run" it all for one year. Once she left, I saw that the pipeline was dry again. I had been duped yet again.

As the CEO, you cannot leave the revenue up to any one person. When they get in your face and say, "I've got this," ask for the data until you feel comfortable. It is your company. Do not feel bad about wanting to see the metrics that you know will ultimately lead to closed sales.

A New Product Emerges

Meanwhile, we spent the next eighteen months from the start of the COVID shutdown working on and launching a new product. We were very strategic in doing that during the downtime, but it also meant having the confidence to invest the money during a pandemic when things were less certain and our revenue was pretty flat.

Our new CTO made an immediate decision to bring much of our infrastructure and development in-house, and worked with a consulting firm he had worked with in the past for the other part. We made this six-month transition off the original company I had selected at the start, and we moved our servers and began bringing development in-house riding the COVID wave.

This would make us more scalable long-term, and we accomplished it during a time when we had fewer new clients going through the implementation process. But it was hugely expensive and certainly not for the faint of heart. We were spending money at a much faster rate than we had during the good times. We focused on cleaning up things we never had time to work on before. We scrubbed the product road map and evaluated future new modules beyond the one we were building.

We began working on automating where we could. For example, our regression testing was very manual. This describes the testing that has to occur when you are going to make a change to your software. It is important to confirm you haven't broken anything else in the software. Every time we did a release, we had to go through testing of each branch of the software manually. It was taking our team three or four full days to do the regression testing; we needed to automate this.

During the early stages of COVID a fellow CEO laid off nine people at her company. Eighteen months later, she was having a hard time rehiring, as was most of the country. As an entrepreneur, you do what you have to do so that your company survives. Sometimes you have to make the tough calls to lay off or not. She did not pause and came out brilliantly on the other side.

Instead of focusing on layoffs, I used the time to double down on investments in the business and decided to move IT in-house and to expand the product. While many of my colleagues hunkered down and didn't spend, we had a base of business to survive on. Instead of investing in securing new sales, we doubled down on investment in the product and our ability to scale more quickly. While there were many ways to navigate the rough waters, I am pleased with our path.

Hospitals hit the financial skids after the CARES Act money dried up in 2022. It was hard to find nurses and doctors, and temporary staffing was taking a bite out of their revenue. The staffing shortage drove the costs of nursing through the roof. The financial stress on the hospitals was coming from all sides. It was also having an impact on our business. I thought after a year or so the hospitals would be investing in technology again, but it seemed the time frame for our prospects to buy would again be delayed. It seemed we just could not catch a break.

The Great Resignation

The headwinds during COVID made it very difficult to manage a company. We had been in a position of trying to scale for the last few years, and the incredible financial pressures the hospitals were under resulted in their decisions being delayed. While we were able to show our software had a promising return on investment, hospitals were unable to make purchases at this time.

The staffing shortage hospitals had experienced had been in part driven by COVID. In addition, though, some clinicians decided, "Hey, it is a good time to retire early." Others left the profession altogether, citing too much stress. At the start of the pandemic, health care workers were celebrated. Over time, they became vilified, as the political and health care environments became much too intertwined. Suddenly health care workers were being attacked inside and outside hospitals. Outsiders denied there was even a virus, creating an impossible situation for workers trying to do their jobs and save lives.

As someone who had spent my life in health care, both in and around hospitals, I was astounded that people began to demand specific treatments that had no basis in scientific fact. It was as if everyone, no matter their professional training, suddenly was a physician. I was shocked at how many people had strong opinions on vaccines and that a world health crisis had suddenly become a political discussion.

In the summer and fall of 2021, with the world not quite over the pandemic, another challenge emerged, which involved a whole lot of people resigning! Yet another headwind in leading a company had arrived: the Great Resignation. As it turned out, COVID had a significant impact on how people saw their jobs and their lives. Many people were ready for a change. Many had decided they appreciated the freedom in working from home. Some decided, "Hey, life is short. I'll work for six months, then enjoy life for six months and simply get a job on the other side." And they could!

I graduated from college in 1988, at a challenging economic time in the United States. I could not find a job in my chosen field and had to take a job as a bank teller. There simply were no jobs for college graduates. I opted to go work for a cruise line to save money for graduate school in the hopes that the economy would be on track and my additional education would be helpful in getting me into the profession I had studied for.

When I graduated in 1992 with two graduate degrees, the job market was still awful. With my generation, the Gen Xers, we are scared to death of not having a job. Simply quitting would never be an option. Yet some during COVID were doing just that.

During the first eight years of my company's history, we literally had two people resign. Now, the number who had left in a nonvoluntary fashion was much higher, but only two had ever left. In September of 2021 I had four people resign. Four out of twenty-four, or 17 percent of my employees. Terrific.

Where Are All the Workers?

How to replace all these people now became paramount. We began our standard networking processes, posting the open positions, but something else had changed. It was basic economics: the supply of people wanting to work was much less than the demand for qualified workers in our very specific health care niche market.

What happens then when demand for workers is greater than the supply? Well, wages go up. Employers get really creative with ways to attract new workers, who are most likely already working somewhere since the demand is greater than the supply. Suddenly, new employees' expectations for salary and benefits skyrocketed.

I also noticed that the supply of immigrants seemed to be nonexistent. Because we are a technology company, we hire people with science, technology, engineering, and math backgrounds (STEM). Because many who elect to study in these areas come from outside the United States, there is usually a large number of US universities graduating students eager to stay here. This supply seemed to be reduced to a small trickle during this time as well.

We had successfully recruited two immigrants to Ludi before the pandemic and had worked to secure their H1B visas, having caught them directly out of their studies. This was a great pathway for us to find talent. It required a bit of extra time and money, but it was a great way to secure talent.

It also became a challenge during COVID to retain existing employees. Everyone had been through a hard couple of years and was restless. In some ways, all I had worked so hard to build seemed to be falling through my fingers like sand. One of my employees was offered a job 60 percent higher than we offered, so she left. Would I like to increase everyone's salary by 50 to 100 percent? Why, sure I would. But it was not practical.

We tried to adjust salaries as best we could, to stay in market, but we were coming off a very flat couple of years. The leadership team I'd pulled together made very thoughtful promotions of staff, created career pathways with each member of their staff, and really tried to listen and engage with employees. Many of our team really believed in the vision, knew that their stock options would be worth something in the end, and believed in our solution and stayed.

Emotional Roller Coaster

It was 2021. I, on the other hand, as the leader, had all kinds of things running through my head. It was a very stressful time. As we rounded our ninth year of being in business and began our tenth year, I began to realize how tired I was becoming. When you are the leader, you must be the eternal optimist, and it is sometimes hard to put on that façade every day.

I had always been a super positive person. I was happy as a baby and still make the decision daily to be positive and look for the good. Glass-half-full kind of approach! We had made it through COVID and the Great Resignation; now I was personally ready to move on. I turned fifty-four that year and had had a vision of building this company for ten years and then moving on to my final work chapter. I was ready to do something different.

I decided perhaps it was time to think about selling. My vision had been to grow the company to $10 million in annual recurring revenue (ARR). But my inner voice said to begin exploring options.

One of the smartest hires I ever made in my life was a young lady who was a former investment banker who had also started and run her own company. Her name is Danielle O'Rourke. She is by far one of the smartest people I have ever had the pleasure of meeting. Having a background in investing in companies, she had worked on roughly a hundred deals. She had evaluated companies and knew all the ins and outs of that process—one I knew about from studying but had never gone through personally.

I'll digress for a quick story. I wanted her to be part of the company so that when I was ready to sell, I had an insider who knew more about mergers and acquisitions than I did. She could help me get the company ready for sale in order to get the highest number since she knew more about what drives prices. It seemed a brilliant strategy to me; though we were far off from selling, it felt like a really cool plan to bring her onboard.

We had coffee in the summer of 2019, and great news, she was not feeling challenged in her work and not enjoying her job. Well, bad news for her, but great news for Ludi. As she talked about the challenges she was facing, I literally heard nothing because I was thinking as fast as I could about how to entice her to work with me. I sat there thinking, *How do I get her to come and work for Ludi? What will make it attractive enough?*

Because we were not ready to sell in 2019, I needed a role for her while she was helping me get the company cleaned up and ready to sell someday. So I asked her, "What do you want to do? What job do you want, and I'll create it." I think I shocked her at that coffee meeting; she thought we were networking, which we were!

We settled on a title and position as chief operating officer, as she wanted to work in operations, helping to streamline and standardize where possible. She would help me with finance, as I was still leading HR, finance, product, and IT myself, in addition to being CEO. There were many things she could help with.

At the time she joined, the reaction to her joining was something out of a movie. One of our executives seemed jealous of her joining, and the rest of the ranks were not as welcoming as they normally were with a new employee. I couldn't figure out if I had a "mean girl culture" or if it was something else. I don't know if maybe the team didn't think her role was necessary, but I sure did need some things taken off my plate. To this day, I still cannot figure out what I could have done to make her entrance to the company smoother.

Luckily, she was a very strong person and was not to be deterred. She also is the only person I've ever worked with who was 100 percent confidential. I mean, this lady would take things to her grave. She quickly became a very important part of the team and worked like a crazy person. I've never seen anyone churn out so much good work.

She evaluated Ludi her first month as though she were considering purchasing the company. She identified a list of thirty-six items we needed to address to clean up Ludi for sale. Not that things were bad; these were just ways the company could be further "derisked" for a potential buyer. Some things could be checked off more easily, e.g., begin having formal audits of the full financials. This isn't hard, just a bit time-consuming and expensive. Other items were much larger, e.g., have the software code base we had built audited from the outside to understand any risks and close them. That was the item that led to additional hiring and various structural changes for us.

Danielle joined a year before the pandemic began, and we immediately started working on the list of things to get the company in great shape. By 2021, all the things we had been working on in tandem for the past three years were nearly complete. Everything we worked on together was completely kept between the two of us.

It was time to move forward with selling the company.

Decision Made, Now What?

In the spring of 2021, a year into COVID, with all these headwinds and the cleanup mostly complete, I was ready to start having conversations to sell. Now that I saw the finish line in my vision, I could focus on how to go about selling the company.

Danielle presented to me as though I was an investor, all the options in the market against our current score card. She updated our positives, meaning the things we had in our favor for a good sale and the areas in which we would get "dinged," meaning areas that would drive our purchase price down. She did an analysis of how the financial markets would evaluate us.

We went through this full analysis every year from the time Danielle joined the company in 2019. It included a full SWOT (strengths, weaknesses, opportunity, and threat) analysis of the business. It was fun to see each year what we had checked off our list of things to improve. We dusted it off and got to work again. By the time 2021 rolled around, I started thinking maybe it is time.

One direction we explored was to hire a banker to represent the company and open it up for sale. A banker comes in and helps you package up and put a bow around your company. They manage the whole process of reaching out to all potential buyers in the market at large, in a process that takes everything about your company and the market in which you reside and puts together your story.

The banker might then open the process up to hundreds of companies who are acquisitive. The banker saves you time and effort in managing the list of potential buyers and in courting them. They also take a piece of the purchase price, much like a real estate commission.

At this stage I narrowed our approach to potential strategic partners first, i.e., business partnerships, to see where that took us. We still had growth as our number one priority, so by working with partner companies, we might be able to identify a way to grow our business before going officially on the market. We would be open to discussions and see where trying to grow our business would lead us.

Partner Discussions

We made a list of all the companies in the market we liked that were somehow in an adjunct business to ours. We landed on about thirty-six carefully curated companies to start with. Danielle took the lead on reaching out to these companies for partnership discussions.

This outreach looks a lot like regular sales. You identify potential prospects, then reach out, schedule calls if you are lucky, then try to figure out how to navigate the prospective partner company. For example, are we talking to the right people? Do they see the fit between what we do and what they do? Do they have clients who may need our services? Do we have clients that may need theirs?

I quickly learned this would not be a fast process. I guess I thought end to end maybe in six months we would have some good partnerships in place that could lead to a deeper partnership later. It started to feel like it would be years, not months. This would be a very long process of outreach, following up, discussions, dead ends, re-energized discussions, potential partnership structure discussions, and maybe ultimately a decision to do some work together.

But I also noticed something else: when we reached out to a potential partner, somewhere around three to four months in on the discussion, they would say, "Hey, we should just buy you." But we weren't really ready for that yet; we wanted a partnership in place to help us not only grow but learn more about the other company. We were trying to get business first, then explore later, once we knew more about each other, but it was really hard to keep focused on the partnership first, arrangement second.

We were successful in partnering with a few companies and even getting some sales leads through the process. Sometimes it included us doing a joint webinar, including each other's prospect lists as we talked about health care topics adjunct to our businesses. Other times we would jointly present at a health care conference. The options were endless to explore, but they all took time and never moved very fast.

Until one of them started to get wings. One of the companies we reached out to was very well known in the hospital world and had a great reputation. Their business was close enough to be comfortable and not at all duplicative to ours. They were a service business, and we were a SaaS business, so there was great potential for both of us.

We started to get more serious in discussions with this company, a company that was, in fact, private equity backed. While the discussions were going slowly, they were going. There was great synergy between the content of what each of our companies did. My one hesitation was, this company was a service-based company, and we are a SaaS company. There are inherent differences in how these types of companies work, which was both a negative and also a positive. We would be their first SaaS acquisition, then in the future perhaps the basis for bolting on more SaaS purchases. We embarked on what was to be a solid year process, which seemed good until it wasn't.

The Wash Cycle

After one year of discussions, I was mentally ready for my next chapter. I had been postponing things, business decisions, because I felt the sale was imminent. It seemed like this deal, like others I had worked on, was just a matter of getting across the finish line. We were nearly seven months into our discussions with this one potential acquirer. We had agreed upon the high-level business terms, had signed a letter of intent, and were closing in on the forty-five-day due diligence period when yet another snafu occurred.

The analogy I would use for the discussions about the acquisition of Ludi is similar to when you are getting ready to sell your house. You think about it and begin to form a list of what you need to do to get ready. Someone rings your doorbell and says, "Hey, are you interested in selling your home? We really want to be in your neighborhood. We realize you aren't quite ready to sell, but we have vision."

You say, "No, not yet. I'm cooking shrimp and the place smells, the kids' bedrooms are not picked up, we need to clean out the lower level, paint a few things and clean out the garage, clean up the landscaping."

They say, "No worries, we have vision; we can see beyond those things."

This deal was not going down as I had thought or planned, so I walked away on day forty-four of the due diligence process. I felt like I was on a roller coaster. You know that feeling when you are coming into the station, and it was a great ride, but you are ready to get off the coaster? Well, not yet, sister; buckle up, we are heading out for another round.

It was emotionally taxing. I was exhausted, sad, stunned, tired, and had to get ready for an all-company meeting in Washington, DC, for the next three days. I needed a moment to wrap my head around the change of course. I thought I was nearing the handoff of my company, but it was time to dig deep, find that excitement, and get the company moving faster. I truly enjoyed my role as CEO, so this was just a return to that focus.

I had to get on a plane and then kick off a three-day company meeting where I would lay out the vision for the company over the next five years. I was like an actress who deserved an academy award, because personally I had not had time to process the change.

Vision for the Future

Our company meetings in DC were terrific. The team was so happy to be back together after two years of COVID. My marketing head and my assistant had planned the most epic three days. We had fancy dinners, ghost tours, monument tours—after all, it was DC. We all left excited to be part of this thing called Ludi.

Though we had experienced another loss a few weeks prior of a senior executive where things had not worked out, we had an interim leader and were starting to assess what we had and what needed to be changed in the product and IT departments.

I was knee-deep in a fellowship program with the Nashville Healthcare Council, an amazing program I had been accepted into. It was a great honor to be part of such a group. It was taking a good bit of time, including two full days a month for five months. In between, there were meetings, networking, all amazing, but on top of the due diligence process I was working on and the staff challenges, it was a lot.

I had to dive back into details I was postponing because I was counting on the sale happening: ongoing issues with our human resources vendor that required more time than I had and an out-sourced IT company we worked with to provision our devices that needed to be replaced. These things had to be dealt with and fast. I had assumed some of these support services would be part of our merged company.

All I did was go from land mine to land mine, but it was exhila-rating. I was firing on all cylinders again. I was working heavily on sales, since our leader and main salesperson left in January and February of 2022. We had our best quarter, adding as many new clients in one quarter as we had in the last two years combined. I really did love sales; that part was going well and is what I love to do. But I was exhausted and stressed.

I was personally getting depressed. I needed time to process the deal not working out. For the past six months, I had spent at least twenty-five hours a week in addition to my normal job hours, pulling data, answering questions, rerunning data, answering more questions.

Plus, it was invasive. They asked questions about everything. Why did you make that decision? What were you thinking? Most of the time there was a clear answer, but sometimes I thought, *Yeah, why had I made that decision?*

This on top of at least an already full plate meant that Danielle and I had not really had a weekend off in five months. I needed time to turn things back around, including my positive attitude. I had not had a week off in probably a year, if not a year and a half. I needed an intervention. I booked myself at a spa for a week in Austin, Texas, and then we got down to business.

Trail Markers:
- *Things outside your control will happen to the business; e.g., a pandemic is outside your control. Always being prepared to pivot is the key to moving through these unexpected times.*
- *During COVID, I doubled down on investing in the product, bringing development functions in-house and expanding our product suite.*
- *COVID was a double whammy, creating an environment where hospitals didn't make software investments for two years, thereby hurting our business. The Great Resignation tail on the back side then created an increase in staffing costs, again hurting our business.*
- *Once the business recovered and we made it through COVID, I began to suspect it was time for me to sell the business. Partnership discussions can uncover potential buyers of your business.*

Chapter 12

The Spin Cycle

I was feeling just like your washing machine, the way the agitator moves the clothing one direction then the other. Water fills up, soap goes in, the agitation starts, then rinse, spin, rinse, spin. Some days, I felt like I was in the spin cycle nearing the end, only to have another flood of water and soap inserted and head back to the agitation cycle.

This company was my baby, we were doing well, and though the first acquisition deal had failed, I knew we were a great company. It was time for my focus to become razor sharp again. The leader cannot lead halfway, and I believe others feel our energy. I meditated and focused on getting back in the game.

I focused the company in three areas: sales, bringing more IT in-house, and feature development. Our fourth product was extremely complex and needed to be built out more based on client experience and what we were learning in the market. There was so much opportunity; we needed to be in a position to be able to grab it, and fast.

We spent the remaining three quarters of 2022 very focused. Danielle took the sales organization in addition to her other duties, and I continued to work with IT and product, as the one translating market needs for the product team, and on the structure of IT. We continued as a lean machine and had a good year in 2022.

The first week in 2023 brought our annual sit-down meeting again, and Danielle helped me think through the timing and options for selling. We reviewed this every year, and I decided it was time to take the company to market.

Selling your company to a strategic buyer seems to be the dream of most entrepreneurs, and it was mine as well. This means finding a buyer that has a company that aligns in the market with yours in some way. It might be a company in an adjunct market, one that is in a similar business in a different industry or could be a competitor. The competitor option kind of made me nervous based on how new our market was in the evolution stage.

We had really built a new category of product, and there were now competitors who had entered the market as "me too" options for our first product. We were industry leaders in our area, and the market was a large one that was not yet saturated.

Another approach to selling your company is to recapitalize, taking on more capital to grow faster but stay involved, hoping for a larger transaction down the road. There is a transaction with a private equity firm or another institutional capital partner where the capital table is redistributed. The current owners essentially sell their shares to the capital partner, who has a new ownership structure.

Typically, all small partners and former employees would be cashed out, and those who have put money into the transaction roll forward on to the new capital table. The founder(s) and key management would typically roll part of their proceeds from the transaction into the new company. The founders would usually take a cash payout and become, on a go-forward basis, much smaller owners in the new company. So you are taking "chips" off the table, or said another way, you are getting some cash out of the business. But you are also losing the autonomy you enjoyed prior.

The goal of a financial partner recap is to grow the business aggressively, where the capital partners would be helpful in reselling the company again in three to seven years. It is often called "taking the second bite of the apple", meaning if you rolled some of your ownership into the stock of the new company, you would get another cash event down the road. This type of sale would be to a private equity buyer, and I had PTSD from our 2022 experience, so this option was less my preference.

I had taken considerable personal and financial risk in the early years and was more conservative by nature. But I was open to what would show up as we launched our process. While I did not want to lead the company moving forward after selling, the company needed a good leader to take it to the next multiple of revenue. My evaluation came down to what the best home would be for my company and what I would need to live comfortably moving forward and get a good return for the eleven years I had been working on the company.

At this stage, I was pretty sure a strategic buyer would be our path. We had a great list of those that were great candidates from the work we had started a few years prior. But I was also intrigued with the recap idea, thinking perhaps that was the right thing for Ludi long-term. The market size was so large, and an investment to scale more quickly could be the right thing. I was going to be open about it.

Personal Life Doesn't Stop

At the same time, there were more than a few things at play. First, my mom had gotten quite ill in the third quarter of 2022. After spending seven days in the hospital and a month in assisted living with her, I was exhausted. For eight weeks, I flew or drove every week to St. Louis, spending four to five days there each time, then returning to Nashville for a day or two each week.

I was helping my dad with cooking and coordinating rides for him to visit my mom daily. I had spreadsheets of who was covering what and when. Each evening, I spoke with her clinicians and confirmed the next day of rides. The assisted living was in a part of town that my dad was not familiar with, so we all felt better with that safety net. It was time to have conversations about easing back on the driving, and those were not going well. I was the coach for my mom's care, navigating her diagnosis and the panel of doctors and nurses providing care. It involved daily conversations, arranging appointments, speaking with caregivers, etc.

If you have ever found yourself in this position, it is quite overwhelming. Having spent my career in health care, I didn't know it would be this way. You truly have to fend for yourself. The first night in the assisted living facility, they would not let us bring in my mom's medications; they said it was their policy to administer the medications themselves. I understand why—they needed to be sure the medications as written were being followed—but it also created a mess for us in getting her medications that first day.

I flew back from my fall company meeting in Nashville that I had largely missed and went directly to the facility. My mom was confused, wanted to go home, but needed the extra care. It was eight thirty in the evening, and she said she had not had her evening medications, so I went on a quest to discover why not. The floor nurse, said, "Yeah, the pharmacy did not get them here; I don't know what to tell you."

I called my brother, who brought the huge box of my mom's medications from his home, where I had been sorting them weekly. Medications at this stage in life are tricky, because one of the many doctors on the team is always tweaking something. It turned out to be a mess if I sorted them too far in advance. I smuggled them in and sorted them right there on the floor of her room. This is just one small example of the navigating you have to do for a loved one who needs services. The providers are all too busy to follow the minute details of each person's care path.

That night, I returned to my parents' home, now somewhere nearly midnight, and had a talk with Dad. He could see how hard this was on my brother and me, and on him. He had been resistant to leave his home of thirty years and downsize, but I think at that moment and with my mom's condition, he knew it was time.

My brother and I convinced my parents that it was time to move into the retirement community after my mom was released from assisted living. We had looked at options before Mom got sick, so we knew which community worked best for our family.

It took multiple group conversations, patience, and many tears were exchanged. With dementia diagnoses for both parents, things were going to get complex fast. My brother and I were always in lock-step and on the same page. I cannot imagine the toll this would have taken on me if we were not in agreement. I had heard stories of siblings pitting one parent after the other sibling. I was and am very lucky to have the absolute best brother out there.

We moved my parents into the independent living location they had selected, and we spent six months cleaning out the family home, which was a very emotional and draining time. Because I lived four states away, it was also physically draining for me.

Is It Really Time to Sell?

As I evaluated timing now with a personal reason to boot, the financial markets were in a bit of a flux. The feds were trying to keep a recession at bay, so interest rates were creeping up. There was a significant amount of uncertainty, and the private equity market of investing was slowing. All investors pumped the brakes for the first part of 2023. The company was in a great position as opposed to what had happened in 2016. We had a solid leadership team; the company was moving like a well-oiled machine!

Our customer base, hospitals, had burned all the CARES money, had crazy increases in staffing costs, and were losing providers. This amounted to razor-thin margins for the hospitals, so our growth outlook reflected these realities. It boiled down to more than one person telling me this wasn't the time to sell, but in my gut I knew it was!

I laid out a timeline for 2023. First quarter, hire a banker. Second quarter, spend time getting data room set up and banker getting materials and story ready for market. Third quarter, go to market, get lots of offers. Then fourth quarter, do due diligence and close by the end of the year.

Several years prior, I had thought selling the company ourselves without a banker was an option. My opinion on that changed significantly after having two separate discussions to sell over the years that failed. It always sounds like it will be a good idea. You just find a strategic buyer, they see the value, write you a huge check, and you live happily ever after. While I have heard stories of success on this path, I think it is far more common that it doesn't reap the highest return or best outcome.

Much like selling your home using an agent, I think a banker helps you get a better price. Using the house analogy, they know the market and help you stage your home, encourage improvements only that will benefit the sales price, handle the showing and negotiation for you, and for this they take a commission. After two "for sale by owner" processes, I had experienced a few things that swayed me to hire a professional.

On my own, I would have no leverage. Once I signed a letter of intent, which is a non-binding letter that most commonly puts you into an exclusive period of diligence, you have lost all power when you represent yourself. When the buyer finds something in the deal or about the company they don't like, they come back with a reduced price, often referred to as a re-trade, and you don't have much leverage to negotiate. And for me, the most frustrating loss of leverage was around the timing. There was no urgency. Our second discussion had dragged on and dragged on. We had no leverage. For these two reasons, deal price and timing, I knew hiring a banker was the way to go for me.

Who Is a Banker?

Hearing that word, you probably get a visual of a man in a suit working in a huge bank. Today, this word can mean anyone working in an actual bank or someone who manages mergers and acquisitions. The latter is what we needed, a person to help with the selling side of our company. There are hundreds of companies who might represent companies or represent buyers. I found in our industry it seemed best to start with recommendations.

This was my process for selecting a banker: we tapped into people Danielle knew in Nashville, our venture capital partner made suggestions, and I made a short list of bankers who had been calling on me for years. Danielle was my right-hand person during the sales process, but the decisions were mine as majority owner. So when I say *we*, I mean the two of us, her opinion being held in high regard but the decision being mine. That made many things easier; I did not have to convince my board to sell, since I controlled the board as majority owner. As a result, the selling process was far simplified for me, as it would be my ultimate sole decision.

We interviewed ten bankers, met with six or seven in person, and I was able to narrow it to several Nashville-based organizations. It is a highly personal thing. This banker will help you sell your company if all goes well, so you really have to like them.

This was all pretty emotional for me. It was my company, an important part of my life for the previous eleven years, and I needed the person representing me to really get me and really like the company. Because bankers are highly intelligent, especially concerning financial minutiae, sometimes they lack in personability, even seeming cold. So they really have to like what you do and see the value in it.

Tip 26: Building the Sales Team

I was selling a home in Chicago. I loved this home; it was a beautiful historic brick bungalow. Typical of middle-class neighborhoods in the 1900s, the homes were like a shotgun, meaning laid out one room after the other from street to alley. You could in theory shoot a bullet through the front door, and it would go out the back door. I had owned a pristine period bungalow, and I loved my home.

I hired a realtor to help me sell it. He was the one who was my buy-side realtor ten years earlier. I didn't realize it until later, but I don't think he really liked my home. He wanted me to paint the walls gray, which was super in at the time, re-stain the floors a dark color, and paint the hundred-year-old woodwork white, because that was what people wanted at the time.

Because it was a historic home, the rules applied a bit differently. It took me seven months on the market, two price drops, and finally getting a new realtor who loved this type of home to move it. I switched realtors, and boom, two offers in one week, one at full asking price. I learned about the people who had made the offers. One family had a little three-year-old who the realtor said was a handful. He stuck his hands in the bottom of the BBQ pit, had soot all over them, and started touching things in the home. He terrorized one of my cats, and the parents did nothing.

The other offer was from a couple, no children. He was a research physician at the local university; she was an artist at the high school. They said they were driving by and saw the home, had no intention of moving but loved it that much. I sold them the home, at 2 percent less than the other offer, because they cared. And I cared who bought the home. Selling my company would be similar; I really cared who bought it.

During one of the banker evaluation meetings, I had a weird emotional moment. They were nearing the end of their presentation, and one of the partners asked me for like the fifth time why I was selling the company. I felt like I had been really clear about my reason, so I was leaning toward scratching them off the short list. As I recounted them again, I got choked up. As a female, I hate it when I get teary in business situations. It was real, though, the enormity of what was going on for me, both personally and professionally, as it was all very intertwined.

I knew this was a huge decision, so I reached out to people in my network who had been down this road or those who were somehow involved in and around this business. Sage advice actually came from a friend during a round of golf one Friday evening; he listened to where I was, asked a few questions, and it helped me move forward with my choice.

You have to click in any great relationship, and that is what happened with my banker, Chris Rogers of Zeigler. Chris had years of experience in SaaS health care companies specifically, both buy side and sale side. He was around my age, had a great team, was located in downtown Nashville, and his leadership style seemed to match mine. We met individually a few times after the big formal meeting and got to know each other. Decision made, let's go!

We started meeting with his team to download all the great things about our company. His team did an incredible amount of work, taking all the pieces of data we provided, and began to create our story for sale.

Getting Out to the Market

I am a pretty positive person, as you have probably surmised by now, and I like to have fun. We really got to know their team of four, Chris and his three colleagues, and they, us. In essence, you are paying them, and you are the client. So I really had fun with this.

I said to our bankers, "Look, I am selling my company only once. I was never the prettiest girl at the prom, but whoever has interest in buying us better act like I sure am!" I wanted to make sure potential buyers considered all sides of the purchase, because it mattered greatly to me who the investor would be and that they would see the vision of what the company could be.

I really had fun with them, and as bankers, I don't want to stereotype, but they are really into spreadsheets and all things finance. While I know my way around Excel, I like to kid around, so it frequently caught them off guard. We did most meetings on Zoom, and Danielle and I would laugh when I started on a story and all eyes would look at the screen with the beginnings of a grin. I would make analogies with cats to events we were experiencing. It was fun to make them smile. Plus, we were paying them, so they had to think I was funny.

The first deliverable we needed to refine was the Teaser, a two-page marketing sheet with all the highlights of our company summarized in a sizzling, provocative manner. This is what gets sent out to the list of potential buyers. The bankers needed to digest all the information we had provided through a virtual data room and also hours of face-to-face meetings. They are the ones skilled at knowing what our potential buyers want to hear and how to package it.

It is difficult to describe the amount of data and information we produced over the next six months in this process. The first virtual data room, or VDR, contained roughly twelve folders, each folder representing a functional area—for example, client management, finance, human resources, infrastructure, product, security, insurance, etc. We had a diligence list we started with, collecting files that answered the questions provided by the bankers. Each folder contained files related to that subject matter that would help a buyer understand our business, our clients, our product, etc.

Another huge benefit of using a professional to sell your company is that they know who your buyers could be. They are going to curate a list of potential buyers with you and surface companies you would have never even thought of. This happened with us. We worked on a list of strategic partners we had started creating years ago, mostly those with adjunct businesses. Those with products that also related somehow to physician alignment were all on our list, but then the bankers added more options. The goal was to keep it around seventy-five to one hundred companies in the first pass. This included strategics and a handful of private equity firms.

I was nervous our buyer search would leak out into the market. Of concern were competitors who might hear we were in market and would spin it in a negative fashion to prospective clients.

We were ready with our response: "We are in a wonderful position with our customer base and in the market, and we feel a need to move more quickly now that the market has been proven out. For this reason, we are raising money."

This was a fact; it was time to scale this company, and I did not have the stomach to be the one doing this investment alone, so it was time to partner! It takes money to grow quickly, to invest in the company in all areas at the same time. I was a great CEO up to $10 million in annual recurring revenue, but it was time for a new CEO to take it to $15 million, $20 million, $25 million, and so on.

All this concerned me, but we were pretty successful in managing the very small amount of fallout that actually happened.

Pony Up!

We went to market in July of 2023, with a deadline in August to have all letters of interest returned, these letters non-binding. The way it works is that the banker sends the CIM to potential buyers. If interested, they sign a confidentiality agreement. They then have roughly three to four weeks to review the information and decide if they would be interested in making an offer in the form of a non-binding indication of interest (IOI).

The reason it is nonbinding is that the potential buyer needs to go deeper into the business to confirm their decision. They might find something that would kill the deal, or they might find something that would cause them to want to renegotiate items previously outlined in the LOI. It is a bit of a game really, but the LOI stage means they are really serious about your organization.

The bankers reach out to all of these potential buyers via phone, provide information, answer questions, and work the pipeline of potential buyers. Like any sales cycle, this is only the beginning. Institutional money lenders, meaning private-equity-backed purchasers, respond very quickly. They are fast at returning the confidentiality agreement and also are very interested in knowing who specifically is in market. There is no skin off their back to look at the company; there is no commitment at this point.

The confidentiality agreement must be signed to learn who the company is. Potential strategic buyers are much slower to respond. My hunch is they have fewer "deal people" in-house and are much more choosy in moving forward. They don't want to waste any unnecessary time if they don't feel it is a brilliant fit.

We had great success with 30 percent of the organizations we reached out to, meaning they signed confidentiality agreements to move forward. The next step was for these organizations to dig around in the data room, ask questions, and decide if they were interested in submitting a formal offer letter. Not all would likely submit a formal LOI, but our banker team was very pleased with our response rate.

The LOI would contain specifics of the offer the company would be interested in making to purchase your company. They varied in levels of detail, but all contained a purchase price range, how the cash and equity would flow, what they expected of the CEO and management team and other specifics of their offer.

These letters are not binding, but the next step would be to select only one to move forward with in exclusive discussions.

Narrowing the Field

I will share that I was amazed at the number of LOIs we received. In order to select which buyer we wanted to move forward with, the next step was to learn more about each potential buyer. Our banker did an amazing job of providing a summary for us about each potential buyer. This contained public information, that from the potential buyer and also anecdotal information they knew from being in this space.

We dropped a few potential buyers at this stage, not feeling the fit. We also had companies hearing through the grapevine that we were in market and reaching out to our banker. This included a competitor's private equity firm. They were incensed that we were not interested in a potential merger. I just felt the market for a product I had created a new category for would not be served by our companies merging. I also did not want them to have inside information about us in the event a deal failed, so we declined opening the process to them.

For those that we wanted to move forward with, management meetings were scheduled over a two-week period. There was at least one meeting a day, and there would then be a shared meal, either a lunch or dinner, depending on the time of the meeting. They were scheduled for three or four hours, and we had to prepare a deck of slides to use to discuss the company at large. Often the buyer would send a requested agenda, but they all contained roughly the same items.

These meetings were an opportunity for us to get to know the potential buyer, but more so, for them to understand and ask questions about our business. We worked with the bankers to prepare a deck of slides, and the ending result was just shy of a hundred pages. Danielle and I moved seventeen slides to the front as our executive summary.

Together we worked with the bankers on all the data submissions. She and I would go over files, I would put them into our shared box, and she would review and move them into the VDR for the potential investors to see. Or vice versa: she would put files in the shared box, and I would review and move them. Many files only I had access to, for example, those relating to human resources, product detail, and much of the history of the early days. The development of our story took quite some time, and with questions from those coming for meetings, the VDR grew in size.

These meetings were hard, fun, frustrating; pretty much every emotion you could experience would happen. For me, this company is so personal; it was easy to get frustrated with questions I thought were pejorative. Within five minutes at each meeting, we would sense how much the potential buyer really knew about health care, our space, and our company. Sometimes the delta felt too large, as if a deal with them would be an uphill battle, not to mention the go-forward new company.

These bankers are all finance people by background, wicked smart; a typical profile is a younger, highly intelligent finance professional. You get questions about your days in accounts receivable, AR, your customer churn rate, your revenue, and a thousand other questions around metrics. Some I had heard, and many I had not.

They would ask, "Tell me this, why did you do that?" You feel like you are answering the same questions over and over.

Some are easy to answer, and others make you feel totally exposed. The questions can take on an air of arrogance, as if they know your business better than you or that they know something you don't.

I sometimes wanted to scream, "Hey, what in the heck have *you* ever built? You are making judgments on this process, but it is a relevant business that has traction and is growing!" I reminded myself, they were making a business evaluation, and it was a big decision that needed to be carefully vetted by the buyer as well.

We used our banker's office in Nashville, and all the potential buyers came to us. We developed a consistent routine, with me at the head of the table, Danielle to my right, and Chris to my left. I have very little hearing in my left ear, so placing Chris there ensured none of the potential buyers ended up near my bad ear.

Danielle and I developed a flow, splitting up the executive summary deck. We had great banter, and we had fun with these intense meetings. In terms of the company, only two additional leaders and our assistant knew we were involved in a possible transaction. Keeping the discussions quiet until we had a direction was not easy.

No matter how private we were, little things slipped, meetings on calendars someone in the company saw, etc. Being out for two solid weeks was also difficult to manage internally. The flip side was, it was not helpful in my opinion at the stage we were in for others in the organization to know of a pending transaction. It is a huge distraction for employees. They think, *Will I have a job? What does this mean for me? What will our company look like moving forward?*

At this stage, we had no answers to these questions, so I felt it best to keep this totally under wraps until we knew more about the direction. Our CTO and our VP of client success were also going to be involved in the management meetings, so we brought them into the fold.

I learned there is a whole new world of acronyms and vocabulary bankers use, as in all industries. This jargon included cadence, FOMO (fear of missing out), thesis, tee it up, unpack, plates spinning, whistling past a graveyard, ROI (return on investment), and rationale, to name a few. After the first meeting, I suggested to Danielle we play a game and try to use whatever new word had surfaced in one meeting at the next meeting before the bankers did. When we did, we would smile at each other, laugh, or even give each other a high five or fist bump.

One morning we began the meeting with our third slide, the eight reasons why Ludi was an amazing company, and I said, "Let's double-click on that," one of the phrases from the prior day, and Danielle and I immediately nearly went into hysterics. We could not look at each other. We called this game banker bingo, and at the close of our deal we gave the bankers a real bingo game with all these words!

As a founder, you can be quirky, I learned, and it is your meeting where you are the prize, so I had fun with this. Our bankers had no idea what we were doing and often wondered what was going to come out of my mouth next. Sometimes I would look at Chris and begin a cat story and watch his face nearly melt into panic. He was so afraid I was going to say something inappropriate, but I would go right up to the edge and not cross the line. You have to use humor to survive stressful events, and these meetings sure were.

After answering the same questions ninety-nine times in the same meeting, I would sometimes get snippy. Our most challenging area centered around the company's growth through the pandemic. To me, it was really easy to understand.

There was a pandemic, beginning in Q1 of 2020, called COVID-19. Because people got sick, they went to the hospitals. Hospitals were dealing with a disease they knew little about, it was highly contagious, people initially were dying at staggering rates, and treatment pathways were not yet clear.

During this time, hospitals were not going to make an investment in software. Especially software that was going to be used by front-line workers, the physicians. Our sales really tanked the first two quarters of 2020, as they did for most companies. By June it felt like we were coming out of the slump, and we had three quick sales, deals that had largely been in process before the start of the pandemic.

Hospitals had assistance from the government CARES Act, money that supported them through 2020 and 2021, but it ran out midyear 2022. The financial recovery for hospitals was still in process in 2023, for sure, but it was getting better. But we couldn't erase the two-year gap period that impacted hospitals. During this repeated line of questioning, I could get a bit sarcastic. "I don't know if you heard, but there was a pandemic!"

I just sometimes felt we had answered that question already, over and over. We were essentially asked in different ways, "Why was your growth flat in these years?" Heck, I was *thrilled* to have survived and not laid off any staff, and to say we stayed "flat" meant we had to grow approximately 8 to 10 percent in new business in those years to keep revenue flat!

The Envelope, Please

After a grueling process and two solid weeks attending management meetings and pretending to be on vacation because we were out of the office so much, the decision was made. We felt the best partner would be a private equity firm based in Connecticut that specialized in health care.

The managing partner won the day in terms of packaging a clear LOI and path forward to support in the growth of Ludi. He was concise, clear, eager, and he wanted to get to know us. We began to negotiate the terms of the LOI with him, and I was a bit surprised at how long that took. It was a couple of intense weeks around the deal terms with our law firm, accounting firm, and their attorneys.

We finally settled the terms of the deal and began what I thought would be an easy process of deep due diligence. After all,

I felt they already knew so much about us. Now, anything called "deep" should indicate it will be quite a process. I was naive, thinking, *We had the data room, we met for four hours. Hey, you know all you need to know.* Boy was I naive!

Trail Markers:

- *Taking your company to market on your own, without a banker, can seem an easier way to sell your company, but you may not end up receiving the highest price without a third party who is able to leverage the market and your business.*
- *After focusing in 2022 on getting the business humming again, 2023 would be the year of the sale. I projected it would take a year: one quarter to select a banker, one quarter to get the data prepared, one quarter to go to market, and one quarter to complete due diligence and close the deal.*
- *Selecting a banker is a highly personal process. You need to be convinced that person sees the value in your company, has done significant business in your space, and clicks with you personally.*
- *Mergers and acquisitions is a highly specialized field. In all cases, I recommend business owners hire a banker to help with this work. As the CEO, you have a full-time job already, and it is not selling or buying companies. You need a third party to create the fear of missing out in prospective buyers and to manage the process on your behalf.*
- *The investment you make in the banker should be offset by their ability to secure a higher valuation on your business.*

Chapter 13

The Final Road to Success

We signed the letter of intent at the end of September 2023, and I thought, *Cool, we are right on schedule*. Our LOI indicated we would go exclusive in discussions for forty-five days of due diligence, where they'd look at all the data we had provided, and then we'd paper this deal, check sent, done. Boy was I wrong.

There were nine channels of due diligence. There was the business side they would internally navigate; then after their own review of the other channels—i.e., administrative, finance, human resources, insurance, legal, cyber security, IT, and our product in general, etc.—they would hire a consulting firm in each of these areas. It is not uncommon for companies to spend several million dollars to evaluate an acquisition.

I understood why, but it didn't make the process less invasive. Deal makers might have a long history of bad deals they've analyzed, and once one bad deal happens, they will put safeguards in place never to have another bad deal slip through again.

Take Theranos and Elizabeth Holmes. She went to the institutional markets and said her technology that used one drop of blood to perform hundreds of lab tests worked, and after raising $900 million over several years, the company imploded because the technology did not actually exist. Digging into the technology is a key part of most evaluations.

I understood the importance of the buyer employing various consultants to dig deep into all areas of the business to confirm it was what we said it was. I prepared myself for the onslaught of questions and the invasive review that would be happening soon.

Wait, What, Surgery?

During the early part of 2023 it became evident that the toe surgery I had scheduled when my mom had gotten sick the year prior would have to be done. I am a golfer, and I wanted to make it through most of the season before taking the break. The surgeon said after my toe joint replacement, I would need twelve weeks to fully recover.

Timing such as it was, my surgery was scheduled for the week after we reached our LOI. I figured we would be deep in spreadsheets, but they wanted to come to Nashville for in-person meetings. Probably not unusual, but they were still doing their business diligence to make sure they were keen on the company before hiring all the outside consultants.

It ended up being somewhat difficult to hold them off the week of my surgery; they really wanted to come that week for meetings. I didn't know how hard it would be to navigate crutches, and I knew the day or two after surgery I might still be on pain meds, so having meetings probably wasn't the best idea. Heavens, I might say something inappropriate.

We ended up meeting one week to the day after my surgery, me on crutches with my big foot up on a chair between myself and the managing partner. I was carrying ice bags and hobbling around on crutches, taking Ubers everywhere. It was a sight, but we made it work. Meetings, the dinner after the meeting . . . every interaction featured my foot on the chair!

Wait, Aren't We Friends?

Pretty soon, I came off this really big high where all these organizations were wanting to acquire my company. We'd survived the grueling process of negotiating as many terms as possible in the LOI, and now I was beginning to feel like we were on opposites sides than we were before. The purchaser shifted roles, not necessarily as the one trying to be selected but the one diving in to make sure they were making the right choice in the deal. The whole process took on a new tone at this stage.

It sometimes felt like they were starting from a position of thinking I was hiding some piece of information about the company. Again, I get the detail the buyer needs to delve into, but I am a "what you see is what you get" kind of girl, and that is how I ran my company—from a very high ethical standard. There were no shortcuts in accounting, tax preparation, treatment of employees, etc. I don't think it was unusual for me to feel this shift, but it was nevertheless uncomfortable.

Each of the lines of due diligence were laid out on a timetable. Each one had an accompanying list of items requested. Much like the information technology questionnaires we filled out for our clients, these spreadsheet lists of requests had multiple tabs, often with hundreds of information items being requested. Some were easy to find, e.g., all client contracts, but others were more difficult, e.g., full-year accounting of employee payroll by pay period for the past five years. If you lined up each diligence channel into one big file, there were literally thousands of rows of requests.

We dug into the data collection process right before I went out on surgery, and it continued through the whole process. Danielle would do the first pull of what she could access and send me the items I needed to dig for. She was a data guru, I mean, really fast at all things Excel, Power BI, etc. The amount of data she churned out each evening in addition to performing her full-time job was astounding.

Once the consultant hired to do the channel review for each of the nine data channels had our files, they would then schedule an in-person or Zoom review. We would have several hours together, answering their additional questions; then after this meeting there would be yet another list of requests immediately sent via email. We would answer those, and again, there would be more questions on the questions. And another fresh list of data requests. In most cases we didn't have to do additional calls, but for finance, legal, IT, and product, the calls were very long reviews each with its own project schedule and timetable. It was like running nine implementations at once!

We produced literally hundreds of files, saved to the drives, our banker team helping index all the requests. They would take any new diligence request, e.g., the request from the cyber channel, and pull all files already uploaded. They would highlight for us where the holes were, take all channels of diligence, and run one report of what was still missing each week to ensure we covered every data request.

I had a super smart boss early in my career who said, "Follow the money, and you will understand the behavior."

I understood each of these individual consultants was hired by the private equity firm to confirm their purchasing decision. At times it seemed they would never stop asking questions, because then they would write up the report and it would be over. Plus, they each individually were in essence "signing off" that their review of our company was as expected. Each channel resulted in a formal document back to the purchaser, which later we did get to see. Long story short, it seemed nothing ever ended.

The Hardest Part

While none of this process is easy, I would have to say the most difficult discussion was around the paperwork, the actual legal documents that outlined all the specifics of the deal. The legal review was being managed by a huge New York firm on our little deal.

This firm had a long-standing relationship with our buyer. Getting the legal review started was frustrating; our attorneys and bankers had pushed for this agreement early, but part of the game was holding this review until the buyer was certain about the deal. This is likely the most expensive channel among all review processes.

In hindsight it makes sense that our buyer waited until they had completed their go-or-no-go decision to invest, i.e., the business review, before starting to throw down the real money. For me as the seller, this was incredibly frustrating. I wonder if there is a way to avoid this, and I don't imagine there is.

On one hand, just sending us the draft agreement earlier would have put less pressure on us and our attorneys, but pressure was on their side in this case. The buyer prefers you not to have as much time with the legal documents. They also prefer to drop these after they have seen some of the results of the other channel reviews. It is the nature of these deals, but it is wise to build this into your expectation.

The legal diligence request came in early and on schedule, but we did not get the legal documents, meaning the purchase agreement, until quite late in the process. This is really where the rubber hits the road. There are a handful of things that really matter, and it is hard to sort that out in a hundred-page document, with appendices that are also hundreds of pages.

Our banker team was key here, as they took the first pass and came back with comments. Our law firm, same thing. Then I had to merge what I was hearing from both our banker and attorneys in order to land on how I felt about the specific contract language. This was certainly the first time I was seeing purchase agreements, probably the last I would see, and there was standard industry language at play here. So, the bankers and lawyers I hired were key to helping me make decisions.

Another difficult part was the emotional aspect of this stage. I was selling my baby, and it was incredibly tiring to do this and try to keep up the façade of carrying out day-to-day job tasks. I was still speaking at industry events, running company meetings, searching for staff for open positions, processing payroll, trying to hide the incredible process going on behind the scenes.

Same for Danielle, who was running sales, with only one inside salesperson and one outside salesperson, in addition to her other areas. She was on most of the sales calls, managing the pipeline and trying to close business. It was important we hit our revenue number in 2023, so she was acutely focused on that department in addition to pulling the hundreds of documents, financial statements, etc. for the acquisition process.

The last thirty days before the deal closed, I tried to handle all the deal items and let Danielle run the business itself. After all, she would be the go-forward leader. We were trying to close out the year successfully, preparing the budget for next year, doing employee reviews, navigating the holidays, etc.

She was set to step into the CEO role on the day the transaction would close. I literally had to have a conversation with my banker every afternoon. It was part advising, part counseling, and part venting. There were a couple of big sticking points that could blow up the deal, as there always are.

This was a very hard period—the whole process was, but the last thirty days were particularly chaotic. The diligence requests were still coming in, and I was working close to fourteen hours a day, seven days a week. Emotionally, I was reaching the end of my rope. It felt like I was having a public colonoscopy; they were looking in every nook and cranny and asking questions that always seemed like they were criticizing decisions I had made.

More than once I wanted to say, "Forget it, let's move on to buyer number two!" We had two potential buyers that were very thorough in following up with us through the whole process. One seemed shocked they had not been selected, and the other kept stressing the value they saw in our companies aligning. We had one bird in hand and several in the bush.

The Spotlight

I'm not a person who likes the limelight. People who know me may be surprised by this, but I really don't like to be the focus of attention. I am extroverted and gregarious, but I don't actually like the attention on my personal life.

When my ex-husband and I were planning our wedding, I asked him if he would take one for the team and let my mom run with her vision. She had gotten married when she and my dad were young, and they had a very small ceremony. I really didn't care much about the wedding, nor did my ex-husband.

I was mortified that my mom wanted to invite close to two hundred people to be there to watch us say our vows. Did I love him? Sure, but this seemed to me like a time for privacy. My preference was to run away to Las Vegas and get married by an Elvis impersonator. Some girls, I am told, start planning their weddings when they are young. Sitting down to dream of flower arrangements, color schemes, and the perfect dress never crossed my mind when I was a kid, or a grown-up for that matter. Business plans, yes. Wedding plans, no.

As long as the discussion is about the product or the company, I am fine standing in the gleam of the spotlight. If someone wants to dive into Gail, it makes me uncomfortable. In 2021 I received the Woman of Influence Award in the entrepreneurship category from the *Nashville Business Journal*. I was so excited about the recognition for the company, but I was horrified to learn a photo shoot and a video was part of the deal.

Because I won this award during the COVID pandemic, the video portion was abbreviated, and we each were asked only one question. We had thirty seconds to respond. I used maybe ten seconds of my allotted time. Talking about myself is miles outside of my comfort zone. Writing this book also challenged me to dig deep. For me, it is much easier to write about Ludi, but writing about Gail—now, that is tough!

I did spend considerable time thinking about the implications of this deal and felt like I was as prepared as I could be, but was I?

Preparing for Next Steps

I met Blake Patterson, a vice president at Alliance Bernstein (AB), at an entrepreneur organization event. For him, I was a potential client. For me, I wanted to learn from him the things I should do before I sold my business. I wasn't sure if I had my personal estate set up correctly, and I was aware there were things you could and should do before you sold your business, but they needed to be done prior.

Blake invited me to more than a few sessions at AB where I learned how I could get ready for this ending. I updated all my personal estate paperwork with an attorney. I set up a vehicle for donations in advance. For me, helping people realize their dream of going to school was huge. I don't have kids, and I wanted to begin thinking through how I could set up endowments. I was thankful to have had the right team advising me through this process and set up for the beyond.

I'm philanthropic by nature and was very proud that Ludi donated 5 percent of its profit every year to charity. It is worth learning from others no matter your stage, and I felt that I had things in order, but it took nearly nine months to get everything ticked and tied.

There were also significant tax implications, so I was on the phone frequently with my advisor, Sasan Zamani, from Frazier and Deeter, the accounting firm that had handled Ludi's finances in addition to my personal situation. These two men were critical, alongside Chris, and all three were my most important weapons. They each saw things from a different angle, which became crucial as I evaluated all aspects of the deal being negotiated.

It was time to begin thinking about what I would do after Ludi. Sharing my story was priority one on this list, so thanks for being part of that vision by reading this book. I had learned in my discussions with others who had sold their companies and from the research I had done that the day after the deal closed could be hard.

As the founder of a company, you are singularly focused on the success of your company. Your personal worth, your ego, and the state of your company can all get intertwined. I had learned that giving some thought to your next big audacious goal is important for a soft landing.

I really liked what we do, what Ludi is about. I'd spent my entire career in hospital-physician alignment. I had always worked for a hospital, leading business development initiatives, or for a company doing the same. In all cases, my jobs were focused on how to grow the business, which always involved a relationship of some sort with the doctors who practiced at that hospital.

I don't think hospitals always know how to partner with doctors, but I understood how they tick. I respected how much they have studied to do what they do, how hard they work, and how much they care about their patients. The whole idea for Ludi came from a desire to make the doctor's world a little better. I figured out as I developed our platform what an impact our software could and does have on the hospital's bottom line.

Buying our software was a win-win. The suite of products we built gets to the core of the doctor-hospital relationship, but there is more to do. I kept thinking of new twists on this relationship and more problems to be solved. There were still so many ways we could expand, and I saw a greater vision than when I started.

My family would describe me as a person who works all weekend, but it's different when it's your business. I loved it, so it wasn't work. I do my best work when I have total silence and no constraints on my time. Saturday and Sunday both offer large stretches of time when I can be more focused on things that require thought. But we all need a break, mentally, and that is an area where I am trying to do better.

As I thought about the future, I decided I wanted to do something that would use my skills sharpened from twenty-five years in the business, and I wanted to help others who might be on a similar journey. I wanted to share my story in the hope it would motivate men and women to encourage their daughters, nieces, and neighbors to pursue careers in science, technology, engineering, and math, or STEM. I wanted to have the time to coach other entrepreneurs who were on their journeys. I was beginning to see my future; now I needed to get this deal wrapped up!

Success!

There is a saying in the merger and acquisition world: "Nothing good happens between LOI and deal closing." Said another way, it is in the seller's best interest to get it done quickly. You lose your acute focus on the company; you are distracted.

We were getting really close. We were down to the last few outstanding legal items, and the diligence channels were wrapping one by one. It was incredibly stressful. It was Thanksgiving 2023, and I had been so focused on this process for the past ten weeks, but really the whole year. I had deal fatigue; it was real, but it was a time when Chris, my banker, was key. He would help me understand the options around whatever nuance was being negotiated and what was common practice in deals like this, and he was always good to lighten the mood.

We had a planned closing date, though as it approached, it was apparent that date would slip a bit, but we were at approximately 98 percent confidence it would close. I decided it was time to call the company together and explain what was going on. I had reached out to all former employees who were shareholders already, and I worried about the news leaking out without my ability to present it to the team. It was time to tell our employees the company was being acquired.

We did an all-hands-on-deck call the afternoon of December 7, 2023, the eve of the planned closing date, which would unfortunately slip to the next week. I presented to the team my belief that this recapitalization would really allow Ludi to begin to scale and that it would provide us with a partner that could help us grow.

Because our partner did not have any investments in this area, we were going to be their first company in this space, so there would be future acquisitions that would be accretive to the business. I explained that as the founder I had taken the business as far as I was comfortable. It was time for us to double down, make more investments in the business, and scale this great company we had built.

This was key to emphasize: I had taken my risks early, and I knew it takes a different kind of CEO to take it to the next level. I explained that Danielle would be taking the helm and how the investments being made in the business would take us to new heights.

Everyone was very excited and happy. We had all been part of an amazing team, and now it was time to pour some gas on the fire and grow faster. All of my employees knew the company would someday be acquired, so no one was shocked. Most had some kind of equity, so they would be receiving their stock payout after their options converted on the closing date. Those who were newer, I explained, would receive a transaction bonus. It would be the third bonus I wrote that year. I liked to sprinkle them around like pixie dust; it was fun.

What I remember most is that people did not want to get off the phone. They just wanted to keep chatting. They were asking a few questions, but really, they were complimenting what we had built, together, and expressing their excitement for the future. I got an overwhelming number of calls, texts, even gifts from employees; it was amazing.

I explained that once the deal closed, my role would immediately shift. I was going to cover for an employee out on FMLA who was having a baby, so I would be covering for her, but that was it. The plan was for me to do this for three months, then fade into the deep background, being available for Danielle where ever needed. I would be a board member in the new company, so I was not disappearing. I still would be aligned with the success of Ludi in the future. Now, it was time to dust off the plans for my next phase, with more choices than before.

Trail Markers:
- *Once the LOI (letter of intent) is in place, the work of due diligence begins. This is where the purchaser will dive into all aspects of your business to ensure they want to do the deal, the purchase price is where it needs to be given any risks they uncover, and they understand all aspects of what they are purchasing.*
- *Due diligence requires an intense amount of time and effort on the leadership team of the company to produce and provide all documents, analysis, and data being requested.*
- *Understandably, the purchaser wants to dig into the market for the business and all aspects of the business to make sure they are comfortable with the deal as it stands. Much like an inspection on a new home purchase, this inspection spans all areas of the business.*
- *Finalizing the paperwork on any M&A deal is extremely detailed and time-consuming. There is a give-and-take or a negotiation underway the whole time. This is where your banker will be your savior, handling the tough discussions and managing the sales cycle for you.*
- *The last few weeks were filled with daily changes, discussions, and meetings with the attorneys understanding what the purchaser was proposing and then negotiating what changes I needed for the deal to work for Ludi.*
- *It is easy to focus only on the sale itself, but as the CEO, time needs to be spent contemplating what role you want in the new company and what your personal life will look like after the sale.*

Chapter 14

Full Circle

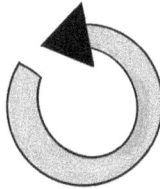

Now that the deal had closed and I was looking forward to defining Gail 3.0, it was nice to take a few minutes and reflect. I firmly believed that hospitals needed what I created and offered—and I still do. I had created a company from nothing. What we sold was the right thing for health care. If all hospitals used it, they would remain compliant, make fewer mistakes, delight physicians, and save money, and it would benefit the industry as a whole.

It was ethically the right thing. We were not selling an experience or a product that was not needed. Tens of thousands of physicians and hospital people were using this software with great success. There was a solid return on this investment. And I felt good about this. I felt good that the purchaser of Ludi was the right partner to scale the business and take it to the next level. I felt good about the team of rock stars that I had built. I felt good that the staff all were able to share in some way in the rewards.

Defining Next Steps

My transition out of my company would happen over a year's time. I would still be on the board of the new company, but I would not be involved in the day-to-day for much longer. I began to think about what my life would look like after Ludi.

When we finally closed the deal, I had been working on this book for years, to commemorate the journey I had been on and all that I'd learned. My dream was that my story might help others with their journey as well. It was time to finish the book and define my goals for Gail 3.0.

I wanted to figure out how to pay it forward and help others on their journey. I began to reach out to organizations and schools so that I might motivate more women and girls to go into STEM fields and careers.

I dove back into my early start-up days and set up a corporation, Gail Peace, LLC, for the release of my book and speaking engagements I might do in the future. I started working on the marketing plan, the website, ThatGailPeace.com, all the little details I had forgotten over the past thirteen years. This time, though, I was fortunate not to have any urgency in timing. I also had the benefit of all that I had learned thus far.

My Legacy

We're all going to leave this planet at some point. I had built a great company and achieved the success I had dared to imagine at the start. I was in charge of my own destiny. How would this sale change me? And would it matter? Was I going to live my life differently in any way? Would I feel left out as I stepped out of the CEO role? Could I find my new purpose?

I pondered these questions as I cruised around my neighborhood in my golf cart to catch the best view of the sunset. The sun was setting over the lake, and a soft breeze was dancing around me. As I watched it slowly disappear into the water, I realized that I'd spent most of my day assisting a friend whose father was in critical condition at the hospital. Another friend was on her way to join me for dinner at the club, which would be followed by another jaunt in Lilly, my lithium-battery-operated golf cart.

We zipped around the neighborhood admiring the beautiful homes, the architecture, and the pretty lights while we caught up on girl talk. Then I texted good night with my new boyfriend. With the sale of Ludi, I had created space in my personal life. I never thought I could be as happy in a relationship as I am at this stage in my life. He is a smart, successful, wonderful, caring man who thinks I am the cat's meow. He is a terrific partner and my biggest fan.

As I crawled into bed and turned out the light, I noticed once again the tattoo on the inside of my right wrist that I proudly wear—the arrowed circle of my company logo. I am honored to have had the opportunity to be part of an amazing company with people I will always consider family.

My Wish for You

Who I am and what I've accomplished look different than I ever imagined or envisioned. Some of life's detours, as I've clearly shown in the previous pages, were not pleasant. But they've brought me to where I am at this moment and have made me who I am today—which I hope is someone of value. Not just someone "with" value.

Whether that means hosting my family and friends in my home for two weeks at a time (shopping, cooking, cleaning, all the while trying to sneak a few hours of work in my upstairs office between meals), offering my condo in Chicago to a friend who really needs a break, encouraging a friend to start her own garden after showing her a bit of mine, spending time with my parents, traveling with my new beau, or loaning one of my cars to a friend whose truck is stuck at the mechanic—that's what makes my heart smile.

What I have isn't here simply to make my life better or easier; it's here ready and waiting to assist those in my world around me. It feels to me like I have come full circle, from glass ceiling to being in charge of my own destiny in the service of others.

I had achieved my goal. I had changed the lives of more than one employee who had received payouts from the sale of the company I had created. Houses were being paid off, vacations realized, college funds stuffed, student loans paid. I was extremely proud of what we did as a super star team. It truly was an American Dream story.

It is not lost on me that we live in one of the greatest countries in the world. I would not have been able to realize this dream in any other place. I am grateful for all I have been able to achieve in this great land.

I survived my crash with the glass ceiling and left to start Ludi. I am incredibly proud of what we built and of the team I assembled and the success we had. It was an amazing journey.

And that's my wish for you. That you'll work hard, make as much money as you're possibly able, invest it wisely, and then lavish it on those you care about and those who are in need— knowing that at the end of the day, it's not really about how much you're worth on paper; it's about being the best version of you, in order to help make the lives of those you work with, play with, and love all the better because you were in them.

The best piece of advice I have is this: You can do it! Bet on yourself. I built a multimillion-dollar company, and you can as well!

Acknowledgments

I want to acknowledge the entire Ludi team, too many to mention by name, present and past, for your dedication and willingness to be on the journey together. You are a team of individual rock stars. You were the best team I have ever had the pleasure of working with in my career, and wish you all the very best success in the future.

I want to acknowledge the many people who inspired and supported me on this journey. To my friend Stephanie, who recognized long before I did, that I wanted to tell my story. To my bestie Karen, who was along for the entirety of the journey—I thank you for giving me honest input on this book.

Thank you to Greg, my ex-husband, whose graphic design for Ludi was timeless, and for the incredible graphics he designed for ThatGailPeace.com, Gail Peace, LLC and That GP Press.

To my partner, Dennis, for your support and belief in me—I am blessed to have you in my life.

About the Author

Gail Peace is a health care executive and entrepreneur with more than twenty-five years of health care industry experience, specifically in the technology space. Gail is the founder of Ludi, Inc., a health care software solution company. Ludi's signature software, DocTime, is Gail's brainchild, a platform she developed in response to a significant yet unmet need she identified during her years as a health system executive. Gail took Ludi from one client to now hundreds of hospital clients nationwide, including some of the largest health systems in America.

Gail is a member of an elite group of females who have raised both venture capital and private equity when less than 2 percent of these investments were awarded to female-led organizations. Gail has been recognized by Becker's Hospital Review as a "Female Health Care IT Leader to Know" multiple times. She was also named one of *Nashville Business Journal*'s "Women of Influence" in 2021 and is part of the Nashville Health Care Council Fellows Class of 2022.

Gail is an accomplished speaker, having presented at conferences all over the country for various health care organizations. She regularly presents at Becker's, the Health Care Compliance Association, the Healthcare Financial Management Association, and the Tennessee Hospital Association. She is also an accomplished author, publishing two to three articles each year in notable trade publications and other resources, including Becker's Hospital Review, Forbes, Healthcare Finance, Healthcare Dive, The Tennessean, and the Health Care Blog.

Gail is today available for speaking engagements through inquiry at ThatGailPeace.com. She advises CEOs on their journey and is active in Nashville, Tennessee, in the entrepreneurial ecosystem. She is involved in paying it forward in the STEM arena as a guest teacher at the high school and college levels. Check out more at ThatGailPeace.com.